Build and Pilot Your Own Walkalong Gliders

Philip Rossoni

McGraw Hill

New York   Chicago   San Francisco
Lisbon   London   Madrid   Mexico City
Milan   New Delhi   San Juan
Seoul   Singapore   Sydney   Toronto

The McGraw·Hill Companies

Cataloging-in-Publication Data is on file with the Library of Congress

McGraw-Hill books are available at special quantity discounts to use as premiums and sales promotions, or for use in corporate training programs. To contact a representative, please e-mail us at bulksales@mcgraw-hill.com.

Build and Pilot Your Own Walkalong Gliders

Photo of Emily McHenry on the front cover used by permission.

Photo credits: dogfight video still in Chapter 7 by Slater Harrison; competition diagrams in Chapter 7 by David Aronstein.

1 2 3 4 5 6 7 8 9 0   DOC DOC   1 0 9 8 7 6 5 4 3 2

ISBN   978-0-07-179055-0
MHID   0-07-179055-1

Sponsoring Editor
Roger Stewart

Copy Editor
Lisa Theobald

Production Supervisor
George Anderson

Editorial Supervisor
Jody McKenzie

Proofreader
Paul Tyler

Composition
TypeWriting

Project Manager
Patricia Wallenburg

Indexer
James Minkin

Art Director, Cover
Jeff Weeks

Acquisitions Coordinator
Molly Wyand

To Chris

Contents

About the Author

Philip Rossoni (Belmont, Massachusetts) is a commercial glider pilot who has always been interested in finding ways to share the experience of piloting an aircraft with a wide audience. He has a Masters of Science degree in Physics from Northeastern University and was at Verizon Laboratories for 13 years as a member of the technical staff, designing light bulbs, electronics, and cellular telephone networks. Philip has also flown as a test pilot, evaluating prototype thermal detection devices for Air Borne Research Associates. He volunteers in the Exhibit Hall Interpretation Department at the Museum of Science, Boston, helping museum visitors learn to fly walkalong gliders. You can contact Phil at info@walkalongglider.info.

Author Philip Rossoni standing in front of a two-person Schweitzer 2-33 of the Greater Boston Soaring Club at the Sterling Airport (3B3). As a commercial glider pilot, he flew demonstration rides for people interested in joining the glider club.

Foreword

My father, Paul MacCready, often spoke of the value of competition for promoting innovation. Among my brothers and me, one form of this competition was to see who could make the best paper airplane. Another of my father's favorite topics was efficiency. In the world of paper airplanes, we saw efficiency in the flattest glide angle. Combining these two influences led us down a path of exploring a vast variety of paper airplanes and other flying creations built from foam, balsa, tissue paper, or anything else we could add nose weight to.

Our most prolific building era was in the mid-1970s. Modern hang gliding was just beginning to catch on, and our father procured bamboo poles and plastic sheeting to make a hang glider that could introduce his sons to flying. When we weren't launching ourselves off the hills, we were making hang glider–shaped paper airplanes.

As hang gliders evolved toward more efficient designs, so did our own creations. We began comparing the efficiency of our designs by gliding them across the living room and seeing how high they would hit the opposite wall. Proceeding without rules, any innovation was fair game. We realized that if the plane flew slowly enough, we had time to run to the far wall and provide a swoosh of air that would make the glider impact a little higher. We did this for a while, wearing ourselves out with a lot of running, until we discovered that we could walk along behind the glider with a board, and, using the same principle as a hang glider or bird soaring on a cliff, we could keep the plane at head height the entire way across the room. Thus was born the walkalong glider.

That was in 1975. The following year, our father began designing a human-powered plane, the Gossamer Condor, and this project gave us access to materials, space, talent, and time, during which we evolved our favorite walkalong glider shape: a swept flying wing with excessive washout (an upward bend of the trailing edge near the wing tip). For us, this design, with its extremely flat glide angle, was the epitome of high performance, and thus it gave us the greatest joy to fly.

In this book, Phil describes several variants on the walkalong glider concept that others have created in subsequent years. Each one highlights a different attribute that has its own appeal, from the grace of the flying wing, to the festive flipping of the tumblewing, to the fascinating but creepy reanimation of the butterfly. Like the Gossamer Condor, these creations are a unique opportunity for a human to power and control an airplane directly. I am sure this book will advance the trajectory of innovation and that more designs will follow.

Tyler MacCready
Pasadena, California

Tyler MacCready is a consultant at AeroVironment, the company founded by his father, Paul MacCready. Tyler was one of the first to pilot his father's human-powered aircraft and appeared in the academy award–winning documentary *The Flight of the Gossamer Condor*, not only piloting the human-powered design but also flying his own walkalong glider design.
Paul B. MacCready, Jr. (September 25, 1925–August 28, 2007) was an aeronautical engineer and designer of several human-powered, human-carrying aircraft, which won the first and second Kramer prizes.

Tyler MacCready piloting his father's creation, the Gossamer Condor, in 1976 while his father, Paul, and a family friend run beside him. (*Photo courtesy of the MacCready family.*)

Preface

The first successful investigations of flight involved airplanes without controls, similar to today's hang gliders. In the 19th century, German flight pioneer Otto Lilienthal would shift his weight from side to side or forward and backward to turn, dive, or climb. In the early 20th century, the Wright Brothers' patent was really about a way of controlling a glider.

Humans took to the air for the first time without controls and, when the very first controls were developed, without an engine. Walkalong gliding takes this same simple approach, where control and power come from you. A walkalong glider rides a wave of air that can also be used to steer it. Walkalong gliding takes something our bodies are not adapted to do—flying—and allows us to power and control a glider using what our bodies *are* adapted to do—walking.

I remember my first encounter with a walkalong glider. Knowing only that it was a glider of some sort, I bought one for my nephew. We didn't know what to make of it, even though the directions were clear. Flying a slow glider with a cardboard paddle seemed foreign, if not unbelievable.

It would take a performance of a walkalong glider flight before I finally understood the concept of walkalong gliding. This opportunity came when the late Dr. Paul MacCready gave a lecture at the Museum of Science, Boston, and demonstrated his son Tyler's Air Surfer walkalong glider. Although I was not fortunate enough to attend the lecture, David Rabkin, currently the Museum's Director for Current Science and Technology, and Michael Scheiss, then manager of Exhibit Hall Interpretation, were present and later introduced me to the concept of walkalong gliding. Their demonstration of the glider was the beginning of my journey, as a volunteer, showing museum visitors how to fly walkalong gliders as part of the Exhibit Hall Interpretation Department.

The last chapter of this book, on ideas for presenting walkalong gliders, is the summary of my experience at the museum helping people learn to fly. Also, as a commercial glider pilot, I believe it's my duty to learn how to fly this new way of piloting models that makes flight accessible to ordinary people, both young and old alike.

I hope this book will guide you on your own personal introduction to the elements of flight, following in the footsteps of the earliest aviation pioneers.

Acknowledgments

I am indebted to Michael Thompson for introducing me to John Collins's tumblewing design. Michael also continues to push the bounds of walkalong glider designs and materials with his wire-sliced Styrofoam materials. Slater Harrison, a schoolteacher from Williamsport, Pennsylvania, has been instrumental in inventing and testing various designs with his classes, organizing walkalong glider meetings, creating the sciencetoymaker.org web site, and introducing me to Roger Stewart of McGraw-Hill, my publisher. Thanks to Connie Hurt of butterfliesandthings.com for supplying me with excellent butterfly specimens. Thanks to Roger and his staff for making this book possible. Thanks to Christina Nicolson of Mrs. Nic's Academia for providing the school enrichment program including walkalong glider instruction. I am grateful to the Westerly Airport Association (WAA) of Westerly, RI and the Experimental Aircraft Association (EAA) Chapter 334 of Groton, CT for hosting a walkalong glider demonstration for several of their EAA Young Eagles Rallies for kids at the Westerly Airport. Special thanks to Gaylene Buck, Anthony DeFrino (then principal), and the faculty of the science department at Paxton Center School, Paxton, MA for hosting me for their professional development workshop in slow flight. Special thanks to Henry Wydler from the Swiss Museum of Transport (Verkehrsmuseum) in Luzern, Switzerland for hosting me on a number of occasions and accepting several early walkalong glider designs into their model aviation archives.

Thanks to my wife, Chris, for putting up with all the airplanes flying around the house.

1

Tumblewing

Project size: Small   **Skill level:** ★☆☆☆

Astrip of paper with four folds is all you need to create this elegant glider design (see Figure 1-1) invented by John Collins, the Paper Airplane Guy. Using a large cardboard sheet (called a paddle) you can steer the tumblewing glider through doorways and perform landings on tables far from where you launch. The tumblewing is the basic trainer walkalong glider and is the best glider design to help you learn how to pilot a walkalong glider.

What you'll need:

- Small sheet of lightweight paper, such as a page from a phone book or a sheet of waxed tissue paper

- Large sheet of cardboard (1 square meter or 3 feet to each side is optimal)

- Ruler (marked in centimeters or inches)

FIGURE 1-1   Completed tumblewing glider

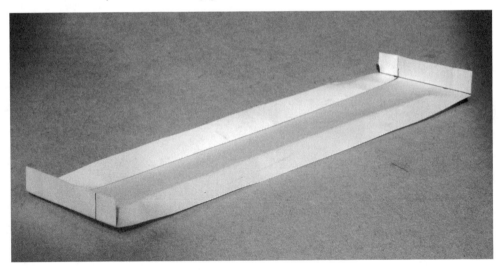

- Pen or pencil

- Scissors

- Box cutter or matte knife

The materials to make the tumblewing paper airplane are all readily available. (You might even find them in the trash or recycle bin.) The tumblewing is best made from lightweight paper such as paper ripped from a phone book or waxed tissue paper like that used in florist shops to wrap flower bouquets. The paddle is made from a large piece of corrugated cardboard cut from the side of a large box, such as a bicycle or large appliance shipping box. The tools you'll need are shown in Figure 1-2.

FIGURE 1-2   Tools you will need to complete the tumblewing glider

Assembly

1. Using a ruler, draw a 5-by-20 cm (2-by-8 in.) rectangle on the lightweight paper, as shown in Figure 1-3.

2. Cut out the rectangle along the bold outer lines, as shown in the figure.

3. Using a ruler, mark the fold lines 1 cm (1/4 in.) in from the edge of the paper strip on the top long edge (side A) and two short edges (sides C and D), as shown in Figure 1-4.

4. Using a ruler, draw a dotted line 1 cm (1/4 in.) up from the bottom of the strip (side B) to indicate that this edge will be folded down instead of up, as shown in Figure 1-3. Flip the strip over and use the ruler to draw a line on the other side of the strip, exactly opposite the dotted line, to mark where you will fold side B, as shown in Figure 1-5.

FIGURE 1-3   Diagram of dimensions of paper rectangle, including fold lines; the folded finished glider is shown at the bottom.

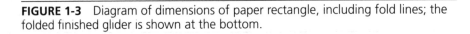

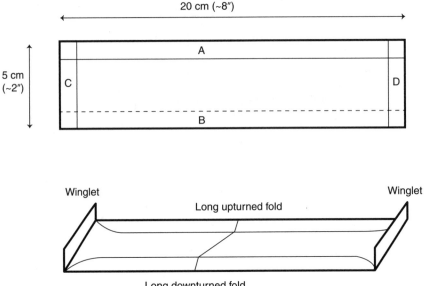

FIGURE 1-4 The strip of paper with fold lines 1 cm in from the edges

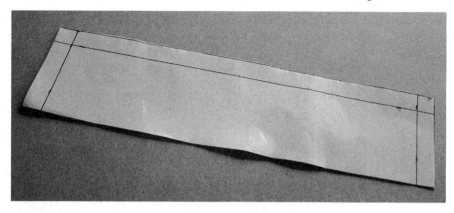

FIGURE 1-5 Draw a line on the back to indicate fold B (as shown in Figure 1-3).

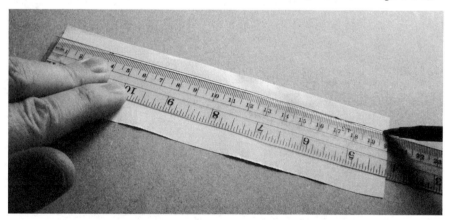

5. Keeping the ruler over the line you just drew, pull the paper up and over the ruler to make the first fold along side B, as shown in Figure 1-6.

6. Flip over the paper and do the same to the solid line you drew on the top, side A, of the paper, as shown in Figure 1-7. The strip of paper should now have both long edges folded, as shown in Figure 1-8, with one long edge folded up and the other folded down.

FIGURE 1-6   Fold the paper using the ruler to create fold B.

FIGURE 1-7   Fold the paper using the ruler to create fold A on the other side.

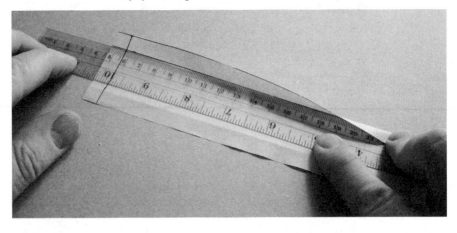

FIGURE 1-8   The paper rectangle with A and B folds completed

7. Now try dropping your tumblewing to see how it flies. Hold the tumblewing by the long edge that's folded *up,* as shown in Figure 1-9. Does it fly sideways sometimes, as shown in the tumblewing flight trajectory on the right in Figure 1-10?

FIGURE 1-9   Launch the tumblewing by holding it by the upturned fold.

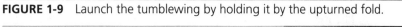
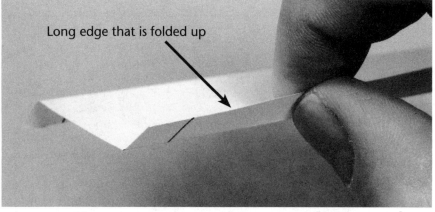

FIGURE 1-10   Comparison of tumblewing flight trajectories with (left) and without (right) the side folds, or winglets (folds C and D, as shown in Figure 1-3), top view

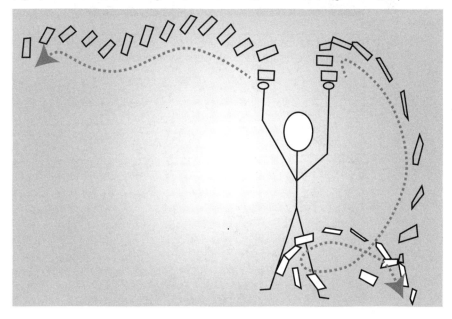

8. Now you'll add the design feature that prevents the tumblewing from flying sideways. Flatten out the long edge folds at the ends of the strip. Use the ruler to fold the short edges (folds C and D), or winglets, of the paper strip up about 90 degrees, as shown in Figure 1-11.

9. The completed fold on each end should look like that shown in Figure 1-12. The completed tumblewing looks like Figure 1-1.

FIGURE 1-11   Use the ruler on the end line to fold each end up by 90 degrees (folds C and D) to create the winglets.

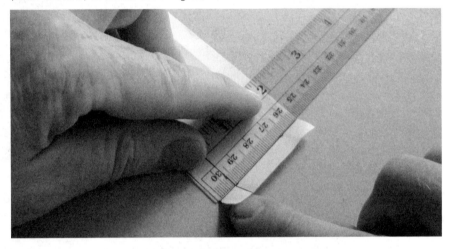

FIGURE 1-12   A completed end fold; note how the folds along the long ends of the rectangle disappear with the new end folds.

Launching the Tumblewing

The tumblewing flies by spinning about an axis horizontal to the direction of flight. In launching the tumblewing, the objective is to get the tumblewing in this spinning flight mode as high as possible with minimum chaotic flight. While spinning, the tumblewing will descend slowest and glide forward, a motion essential for you to create an updraft with the paddle. In chaotic motion, the tumblewing falls straight down. In still air the tumblewing will eventually start flying correctly but by giving it a gentle downward push, the rotation can be started sooner.

1. Hold the tumblewing by the long upturned fold, as shown in Figure 1-13.

2. Give a gentle downward push as you let go of the tumblewing; this will help the glider to start tumbling over and over, as shown in Figure 1-14. The tumblewing should tumble slowly away from you, as shown in Figure 1-15.

FIGURE 1-13   Hold the tumblewing by the long upturned fold.

FIGURE 1-14 As you launch the tumblewing, give it a gentle push down as you let go.

FIGURE 1-15 Trajectory of a gliding tumblewing

3. If the tumblewing does not tumble steadily, and instead stops rotating or falls (as shown in Figure 1-16), try launching in a different location where drafts or wind are less likely to affect its flight (indoors is generally better than outdoors).

FIGURE 1-16   Trajectory of a tumblewing in turbulent air

<u>NOTE</u>   *If you have trouble finding a spot with still air, see "Flying the Friendly Skies (Finding Calm Air)" in Chapter 7.*

Keeping Your Tumblewing in the Air

In this section you will create a paddle and use it to keep the tumblewing in the air and direct its flight. A paddle is simply a large rectangular sheet of cardboard that you use to redirect airflow and affect the flight of your glider.

Using the box cutter or matte knife, cut a large rectangle from a cardboard box, as shown in Figure 1-17. A square meter (about 3 feet to a side) is optimum.

FIGURE 1-17   Use a matte knife or box cutter to cut out a large cardboard rectangle.

Hold the paddle straight up and down, to catch as much air as possible. How does the air respond to the paddle when you move? The air ends up flowing around the paddle, like water in a brook flowing around a rock. The wind created by moving the paddle is also like wind blowing against a building, creating an updraft which birds can use to fly without flapping their wings. Figure 1-18 is a diagram of a bird flying in the updraft of wind blowing against a building.

In the updraft, the closer the bird flies to the building the stronger the updraft will be. In the same way as the bird uses the updraft, we'll make use of the air flowing up and over the top edge of the paddle to keep the tumblewing flying. The only difference is you will be moving the paddle through the still air, rather than the air moving against the building. Once again, the closer the tumblewing flies to the top edge of the paddle, the stronger the updraft will be.

Because the "wind" you will create with the paddle is so small, it is important that you fly your tumblewing in a place with as little wind or

FIGURE 1-18 Diagram of side and front views of wind blowing against a building. A bird exploits the updraft by flying close to the left top corner of the building in the side view, catching the wave of rising air.

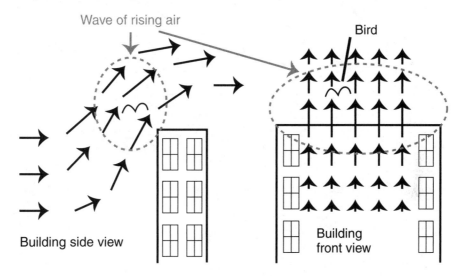

drafts as possible. The small amount of airflow created by the moving paddle is all you need to sustain and control your tumblewing. Trying to fan the tumblewing with the paddle will create too much turbulence, making it all but impossible to keep the tumblewing under control. The motion of the paddle should be smooth and steady, adjusting only to changes in motion of the tumblewing. Think of moving the paddle so as to fly in close formation with the gliding tumblewing. Don't get discouraged if your tumblewing is hard to control—it might be that turbulence is interfering with the flight, not necessarily your piloting technique.

Generating an Updraft with the Paddle

From as high as you can, launch your tumblewing as described in the previous section. It is important that you allow the tumblewing to fly for a short while on its own at the beginning, so it will become stable in flight.

With your paddle in hand at about eye level (see Figure 1-19), start moving steadily toward the glider as it descends. Move at the same speed as the tumblewing to keep the airflow created by the paddle moving around the

FIGURE 1-19 The white arrows show the resulting updraft that sustains the tumblewing in flight.

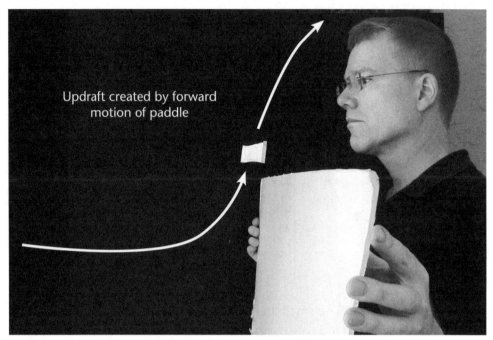

Updraft created by forward
motion of paddle

tumblewing. Think of moving the top edge of the paddle as if flying in close formation with the tumblewing. When the tumblewing is just in front of and above the top edge of the paddle, move the paddle a little faster, as if to run it over the tumblewing. See Figure 1-19 for proper positioning of the paddle relative to the airborne tumblewing. So the tumblewing will determine where to position the paddle, but you can influence the tumblewing's flight with the paddle. This is when magic happens: the paddle creates airflow, and the tumblewing will stop heading downward and may even start flying upward, as shown in Figures 1-19 and 1-20!

Figure 1-21 shows the paddle position for the same tumblewing flight trajectory shown in Figure 1-19. Note how the paddle is brought closer and closer to the tumblewing until it is just about to overtake the tumblewing. In future figures with flight trajectories the paddle will be left out for clarity. See www.walkalongglider.info to view the video clips of each trajectory in the figures in this book.

FIGURE 1-20   Trajectory of tumblewing flight with the paddle generating an updraft and sustaining the tumblewing

FIGURE 1-21   Trajectory of tumblewing flight with the paddle generating an updraft. The paddle has been left in the picture to show its position. To view the video clips from this book, please see www.walkalongglider.info.

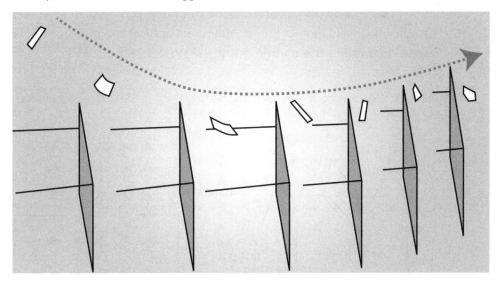

It is important that you hold the paddle vertically to generate the fastest updraft. Figure 1-22 shows how the updraft is affected as the paddle is tilted toward the horizontal position.

FIGURE 1-22   Keep the paddle in the vertical position (left). As the paddle is tilted toward the horizontal (middle), the updraft decreases. In the full horizontal position (right), no updraft is created.

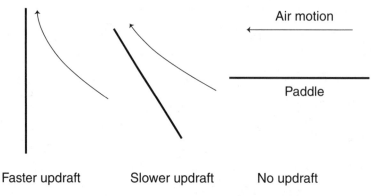

The speed with which you move the paddle is important. The best speed is a balance between too slow, when the tumblewing will keep on descending, and too fast, when the tumblewing either crashes into or flies up over the top edge of the paddle. Optimally, the top edge of the paddle is positioned as close to the tumblewing as possible to surround the tumblewing with a strong updraft. If the paddle is pushed too far ahead, the tumblewing will fall into the turbulent air behind the paddle. See Figure 1-23 for a diagram of the paddle, tumblewing, and air motion.

FIGURE 1-23   Side and front views of a tumblewing flying in the updraft created by the moving paddle.

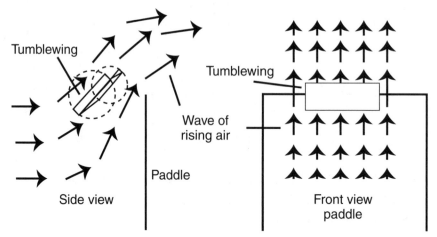

When scientists are doing an experiment, they might take the paddle and tumblewing through a full range of paddle motions to see what happens. So try going a little too fast to see how the tumblewing reacts. Then try moving too slow. Experiment!

<u>*TIP*</u> *Don't get discouraged if the tumblewing doesn't fly well at first. Keep on trying by flying in different places and moving at different speeds, because other drafts may be interfering with the wind you are generating with the paddle.*

Steer Your Tumblewing

Does your tumblewing turn to one side instead of going straight ahead, resulting in a spiral flight trajectory like that shown in Figure 1-24?

FIGURE 1-24 Spiral flight trajectory of a turning tumblewing

Try keeping the paddle facing straight at the tumblewing as you move forward, even while the tumblewing is turning. This should help the glider fly in a straight line or even turn in the opposite direction. See Figure 1-25 for a top view of paddle and tumblewing turns.

By angling the paddle vertically relative to the tumblewing, you can get it to turn in either direction (see Figure 1-26). Remember to hold the

FIGURE 1-25   Top views showing the angled position of the paddle to turn the tumblewing either to the right or left.

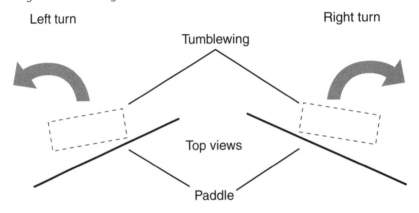

FIGURE 1-26   Angle the paddle relative to the tumblewing to make it turn.

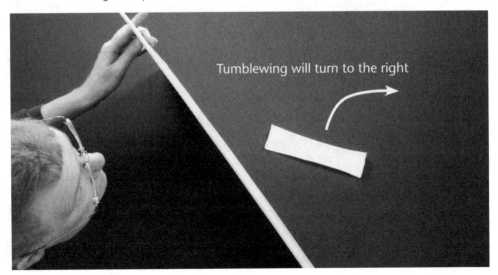

FIGURE 1-27   Flight trajectory of a tumblewing paper airplane performing the S turn maneuver.

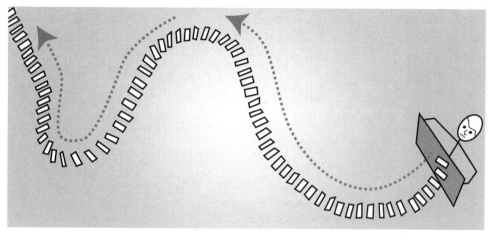

angle relative to the angle of the tumblewing, and keep following the tumblewing as it moves; it may take time to respond. Once you get the hang of this, try performing an S turn, as shown in the flight trajectory in Figure 1-27. Keep following your tumblewing with the paddle to affect its flight path.

Make Your Tumblewing Fly Higher

As soon as you've managed to keep your tumblewing flying as long as you want, try and get it to fly higher. Once again, the speed you move the paddle is important as well as how close the paddle's top edge is to the tumblewing. Getting your glider to fly higher is just like keeping it flying, except you will need to be closer to the "critical speed" and the paddle will need to be even closer to the tumblewing, without crashing into it. As the tumblewing flies higher, raise the paddle along with it to supply an updraft and help the glider gain altitude, as shown in Figure 1-28. Flying higher is safer for any flight and will make it easier for you to help the glider recover from upsets caused by drafts or turbulence.

FIGURE 1-28   Flight trajectory of a climbing tumblewing

Fly Your Tumblewing Indoors

The tumblewing design is the slowest and most maneuverable of the paper airplane designs. This is the best design to try in relatively confined spaces such as inside the home (see Figure 1-29).

FIGURE 1-29   Tumblewing being maneuvered through multiple doorways

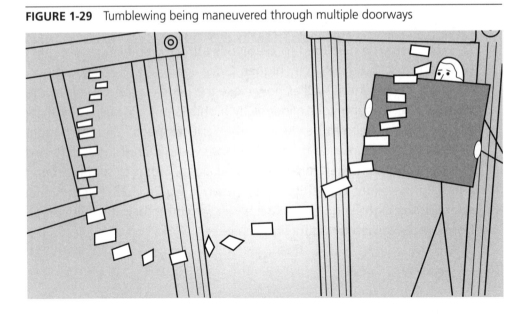

Theory

There is no engine in the tumblewing, so where does the energy that keeps the glider in the air come from? You! Your energy, creating an updraft under the tumblewing with the paddle, is what keeps it flying. When you follow the glider in flight and keep the upper edge of the paddle close to the tumblewing, the air moving up over the paddle rises faster than the glider moves down.

How can the glider be steered when nothing on it can move? How you orient the paddle and push the air with the paddle allows you to steer the tumblewing in flight. To steer the tumblewing, angle the paddle vertically (about the vertical axis, also known as the yaw axis) in the desired direction of flight, which makes one end of the tumblewing closer to the paddle, where the air is moving up faster. This faster moving air pushes one end of the tumblewing up, resulting in a turn.

Controlled flight is more important than sustained flight. After all, what good is an airplane if you can't steer it? A popular misconception about flight is that an engine or flapping wings are required for flight. In fact, most flying things will glide if the wings stop flapping or the engine is shut off. Your ability to fly the tumblewing means that you are doing a controlled flight.

The tumblewing will descend at a constant speed: The weight of the tumblewing is counteracted by the downward acceleration its falling motion imparts to the surrounding air molecules. So making those molecules go up as fast as the tumblewing comes down results in no net downward motion of the tumblewing relative to the ground. This phenomenon is called "soaring flight." The Wright Brothers used soaring flight to fly their glider in October 1903, before they added an engine. In fact, they were able to fly for a longer period in the air rising up the face of the dunes at Kitty Hawk than they did in any of the first motorized tests in December 1903. Just as the Wright Brothers gained experience flying airplanes using soaring flight, you'll get better at using your cardboard paddle and tumblewing the more you practice.

The paddle's job is to produce the upward motion of the air molecules by virtue of its motion through the air. The larger the area of the paddle, the faster the air will rise for a given horizontal speed. So a larger paddle can be like a more powerful engine, making it easier to direct the tumblewing upward. In addition to the air moving up and over the paddle, there is a component of air motion that is counter to the direction of motion—in effect, a headwind, as the air moves backward to fill the lower pressure area behind the paddle. So a larger paddle will produce a larger effective headwind, resulting in slower ground speed for the same tumblewing.

Soaring flight is used by anything that flies to stay up or fly higher. Some species of birds rely almost entirely on soaring flight for survival. For example, heavy, meat-eating birds of prey such as eagles, falcons, and vultures rely heavily on soaring flight when they take wing.

Experiment 1: Wing Loading of Tumblewings

The paper used to make the tumblewing has material properties that affect how fast the tumblewing descends.

- What happens when you make a tumblewing from copier paper instead of waxed tissue paper?

- How much does a given size of paper weigh?

The area (width × length) of the tumblewing affects how much air it catches. If you use a heavier paper, without changing the area of the tumblewing, it will descend faster, because the same amount of air will need to be accelerated more quickly to equal the increased weight.

Figure 1-30 shows two tumblewings made from different weights of paper. The heavier paper at the left is falling faster than the lighter paper tumblewing on the right. The weight of the glider divided by the area of its wings is called the "wing loading."

FIGURE 1-30   Flight paths of two tumblewings made from different weight paper. The heavier paper at left falls faster.

The tumblewing made from heavier paper will need to fly faster, and the paddle will need to produce a faster updraft to keep it from floating down. Correspondingly, different size tumblewings made from the same weight paper will fall at the same speed.

Table 1-1 shows the weight of various types of paper of the same size per square cm, or square in.).

<u>NOTE</u>   *Why do we use the Depron foam since it is so heavy? Foam holds its shape better than paper, especially for larger designs like the Jumbo walkalong glider in Chapter 4.*

TABLE 1-1   Weight of Various Types of Paper of the Same Size

Paper Type	Weight per Unit Area in Milligrams per Square Centimeters	Weight per Unit Area in Milligrams per Square Inch
Waxed tissue paper	2.1 mg/cm²	14 mg/in²
0.6 mm thick sliced foam	3.4 mg/cm²	22 mg/in²
Light newspaper	4.0 mg/cm²	26 mg/in²
Copier paper	7.5 mg/cm²	48 mg/in²
1 mm thick Depron foam	9.0 mg/cm²	58 mg/in²
3 mm thick Depron foam	10.7 mg/cm²	69 mg/in²

How did we make the measurements used in Table 1-1? Figure 1-31 shows the triple-beam balance used to determine the paper's weight. We divide the measured weight by its area in square centimeters to determine the weight per unit area.

How does your favorite paper compare with the papers shown in Table 1-1? You can figure out your paper's weight per unit area by dividing the weight by the dimensions of the paper. So if you have a piece of paper which weighs 3 grams and is of dimensions 2 cm by 5 cm:

3 grams (1000 mg/gram) / (2 cm x 5 cm) = 300 mg/cm²

FIGURE 1-31   A triple beam balance used to measure the mass of a piece of material

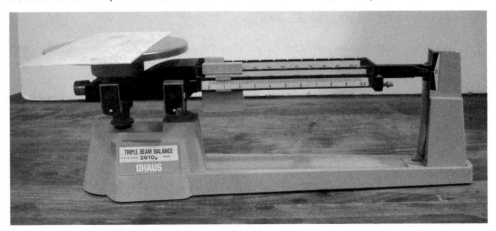

What's the Difference Between Weight and Mass?

Mass is a measurement of how much matter is in an object. Weight is a measurement of how much gravity is pulling on that object; gravity actually affects the weight of the object.

An object's mass is the same wherever it is—on Earth, on the moon, floating in space—because the amount of *stuff* it is made of doesn't change. But the object's weight depends on how much gravity is acting on it at the moment; an object will weigh less on the moon than on Earth (the moon's gravitational pull is about 1/6 as powerful as the Earth's gravitational pull), and in interstellar space the object would weigh almost nothing at all.

Weight is measured using a scale. Mass is measured using a balance (like that shown in Figure 1-31). Scales that use calibrated springs measure this gravitational force and measure the object's weight. Because gravity is more or less the same everywhere on the surface of the Earth, the spring scale can safely be calibrated to determine the object's mass, but only when subject to Earth's gravity! A spring scale will no longer be reading mass if it is used, say, on the moon, which has a gravity of 1/6 that of Earth! On the other hand, the triple beam balance shown in Figure 1-31 would read the correct mass on Earth or on the moon, because both things on either side of the balance (the object being weighed on the left and the weights on the right) are affected by gravity in the same way—no calibration involved.

Experiment 2: Rotation Rate of a Tumblewing

How does the ratio of length to width of the paper strip (known as the "aspect ratio") affect the rotation rate of a falling tumblewing? To determine this, follow these steps:

1. Create two tumblewings in which the ratio of length to width is the same, but the sizes differ between the two. Cut one paper strip with dimensions of 3-by-12 cm (1-by-4 in.) and a second strip of dimensions 6-by-24 cm (2-by-8 in.). Then fold them appropriately.

2. Drop both tumblewings together and note whether one spins
    faster than the other. You can also measure the spin by making a
    video of the falling tumblewings and counting how many frames
    each tumblewing takes to flip. (I used video stills from QuickTime,
    downloadable at www.apple.com/quicktime).

3. Now make two tumblewings of the same width but with different
    aspect ratios. Cut one paper strip of dimensions 5-by-12 cm
    (2-by-5 in.) and another with dimensions of 5-by-25 cm
    (2-by-10 in.) and see which one tumbles faster than the other.
    Is the rotation rate affected by the aspect ratio or the width of the
    paper strip?

2

Paper Airplane Surfer

Project size: Small   **Skill level:** ★★☆☆

The paper airplane surfer, shown in Figure 2-1, is a more traditional airplane than the tumblewing from Chapter 1, because it flies straight and does not spin, or tumble, like the tumblewing. To fly straight like this, the paper airplane surfer needs a nose weight made from folding the paper over on itself multiple times. Because of this, the airplane surfer flies faster than an equivalent tumblewing made from the same weight paper (see Experiment 2 in the Theory section).

Here's what you'll need:

- Small sheet of lightweight paper, such as a page torn from a phone book or a sheet of waxed tissue paper

- Large sheet of cardboard (1 square meter, or 3 feet to each side, is optimal). The cardboard from Chapter 1 may be reused too.

FIGURE 2-1   Completed paper airplane surfer

- Ruler (marked in centimeters or inches)

- Pen or pencil

- Scissors

- Box cutter or matte knife

Assembly

1. Using a ruler, draw and cut a 17-by-22 cm (6 5/8-by-8 1/2 in.) rectangle from the lightweight paper, as shown in Figure 2-2. (If you are using phonebook paper or light newsprint, see the section "Experiment 1: Paper Grain Direction" in the Theory section of this chapter for the best orientation of the longer edge of the rectangle.)

FIGURE 2-2   Start with a rectangular-shaped lightweight paper; here, the superimposed line shows where to make the first fold.

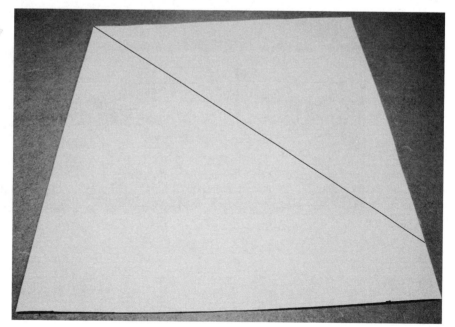

2. Fold the upper-right corner diagonally to align with the left edge, as shown in Figure 2-3. As you can see, you do not fold the entire sheet of paper in half diagonally. Note that aligning the edges results in about 5 cm (1 7/8 in.) below the fold at the bottom of the sheet. Crease the fold.

3. Unfold this crease and do the same with the upper-left corner (see Figure 2-4), folding it diagonally to the right edge, as shown in Figure 2-5. The fold should end at the same place the fold on the other side ended. Crease this fold.

FIGURE 2-3   Fold the upper-right corner diagonally so the top edge aligns with the paper's left edge.

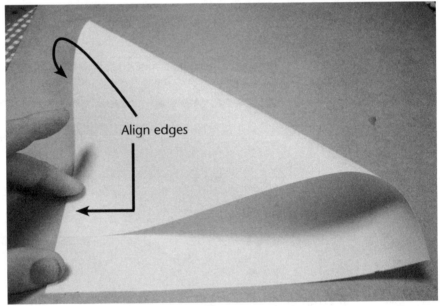

FIGURE 2-4   Fold the upper-left corner to the right edge. The superimposed line shows the position of the next fold.

Align edges

FIGURE 2-5   Fold the upper-left corner diagonally so the top edge aligns with the right edge.

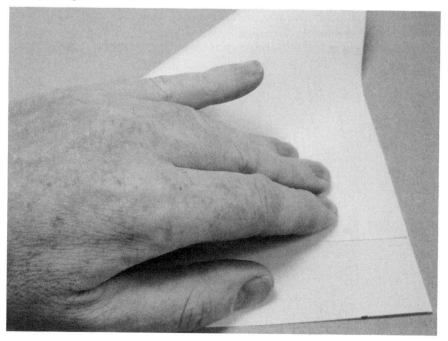

4. Unfold this new crease. The resulting folds should form an "X" (see Figure 2-6).

5. Now pinch in the sides, as shown in Figure 2-7.

FIGURE 2-6   The X-shaped fold

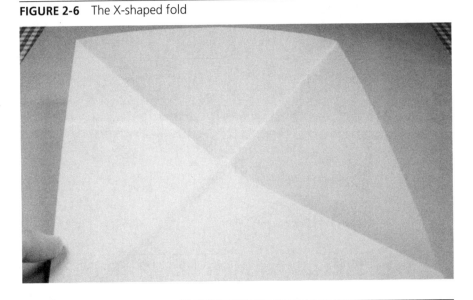

FIGURE 2-7   Pinch in the sides of the X-shaped fold.

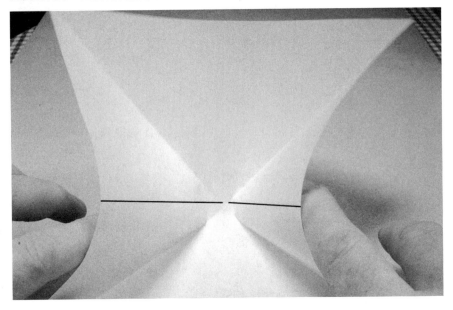

6. Keep the top triangle flat with your hand, as shown in Figure 2-8.

7. Align the right and left corners of the top triangle with the edge of the page, as shown in Figure 2-9. Before creasing the side folds

FIGURE 2-8   As you pinch the sides, keep the top triangle flat.

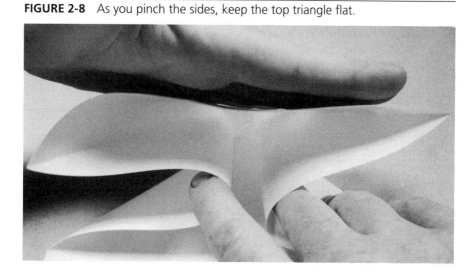

FIGURE 2-9   Align the right and left corners of the top triangle with the edge of the page.

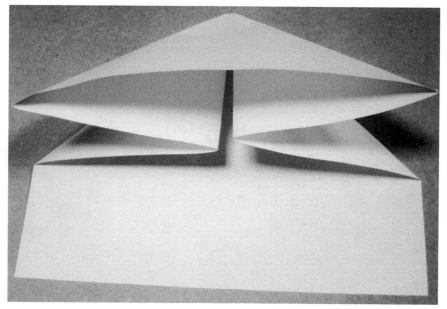

along the centerline, make sure the left and right bottom corners align with the edges. The sides should be sandwiched between the top and bottom layers, as shown in Figure 2-9.

8. Press down in the center and crease the centerline fold toward the top point. All the edges should make a straight line, as shown in Figure 2-10.

9. Fold and crease the right (Figure 2-11) and left flaps (see Figure 2-12) in and up to form a square (Figure 2-13).

FIGURE 2-10 Crease the center fold between the two sides. The superimposed line shows the position of the next fold.

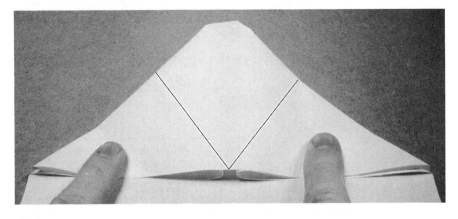

FIGURE 2-11 Fold the right flap up to align with the top corner.

FIGURE 2-12   Fold the left flap up to the top corner.

FIGURE 2-13   Resulting square shape after folding the right and left flaps to the top corner. The superimposed line shows the position of the next fold.

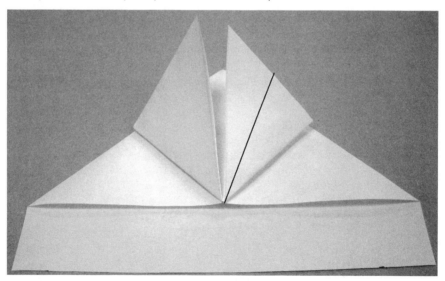

10. Fold both right and left flaps a second time to align with the centerline, as shown in Figures 2-14, 2-15, and 2-16.

FIGURE 2-14 Fold the right flap in toward the centerline.

FIGURE 2-15 The completed right fold. The superimposed line shows the position of the next fold.

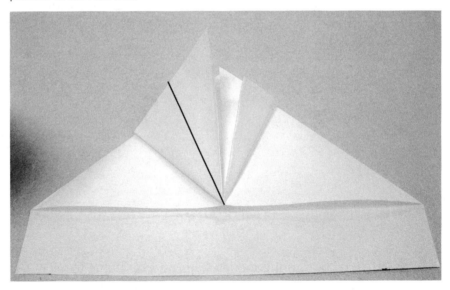

FIGURE 2-16 The left and right folds should look like this when completed.

11. Fold the right and left top flaps down out of the way, as shown in Figure 2-17, so the top of the folded plane looks like Figure 2-18.

12. Open the side pockets of the top corner, as shown in Figure 2-19. This is the nose of the plane.

FIGURE 2-17 Fold the top right flap down.

FIGURE 2-18   Completed top folds

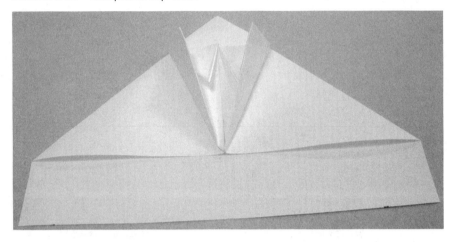

FIGURE 2-19   Open the pockets of the forward nose.

13. Insert the right and left flaps into the opened slots (see Figure 2-20) and flatten the folds. The finished nose should look like Figure 2-21.

14. Turn the paper upside down (this is the top of the plane) and locate the centerline by pressing down on the bottom tip of the fold on the other side, as shown in Figure 2-22. (Alternatively, you can fold the plane in half along the centerline at this time, as shown later in Figure 2-34.)

FIGURE 2-20   Insert the flaps into the pockets to seal the fold.

FIGURE 2-21   The completed nose weight created from the paper being folded multiple times

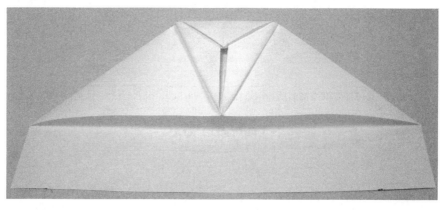

FIGURE 2-22   Locate the centerline by pressing into the back of the folded paper with your finger.

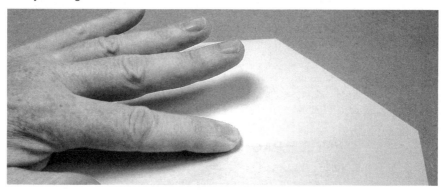

15. Fold the right side toward the centerline, as shown in Figure 2-23. Do the same for the left side. These sides are the plane's wings. The plane should now look like Figure 2-24.

FIGURE 2-23 Fold the side wing to meet the centerline of the plane.

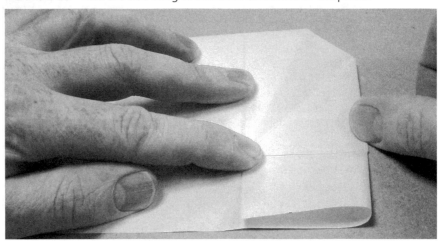

FIGURE 2-24 The completed vertical folds of the right and left wings. The superimposed line shows the positions of the next folds.

16. The next folds will make the wing tips look a little abnormal, but this feature will help you steer the paper airplane surfer with your paddle. Unfold the left wing fold. Then steady the leading edge fold of the right wing fold with your index finger and match the front edge of the wing with the side edge, as shown in Figure 2-25.

17. Keeping your index finger on the forward fold, roll the wing tip outward with your other finger until the wing tip edge meets the corner fold, as shown in Figure 2-26. Do the same for the left wing so the plane looks like Figure 2-27.

FIGURE 2-25   Crease only the forward part of the next fold, aligning the front edge with the side edge.

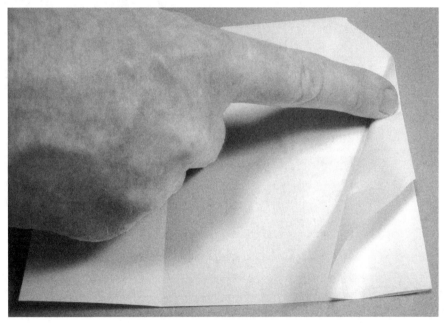

FIGURE 2-26 Roll the wing tip outward to meet the back corner of the side fold.

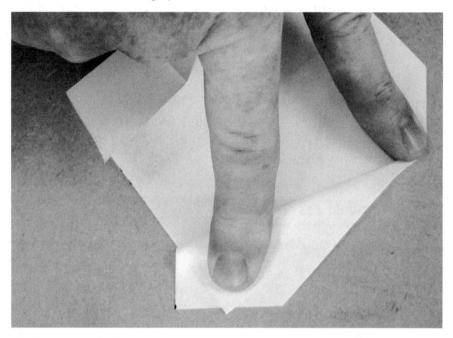

FIGURE 2-27 The completed wing folds

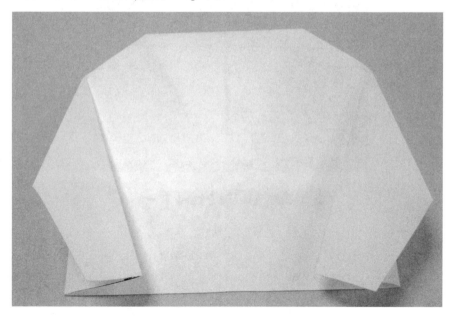

18. Turn over the folded plane to the side you started with; this is the underside of the airplane. Unfold the wings. Using a ruler, draw lines from the trailing edge along the outermost fold and angling in to the centerline, along a line from where the rear end of the folded nose weight ends to each corner, as shown in Figure 2-28.

19. Draw lines along the outermost fold crease to the trailing edge so the lines look like those shown in Figure 2-29. You will be trimming

FIGURE 2-28   Draw lines along the outermost fold to a line from each corner to the centerline just behind the nose weight.

FIGURE 2-29   Lines indicating the area to be cut from the tail

the paper at these lines. Note that the outermost trailing edge is not trimmed. You'll leave the outermost trailing edge untrimmed; they'll be used as trim tabs, or "elevons" (combined elevator and aileron controls).

20. Cut along the lines you just drew, from the trailing edge to the centerline, to remove a triangle-shaped piece from the tail, so the plane looks like Figure 2-30. Removing this tail section will move the center of gravity toward the nose and smooth out the plane's flight path.

21. Flip over the plane horizontally so the cut area is now at the top. Fold in the wings along the creases you made. Place the ruler from the front centerline to the vertical wing fold, as shown in Figure 2-31.

22. Fold the leading edge up along the ruler, as shown in Figure 2-32. Do the same for the left side so the leading edges are turned down (the plane has been turned right side up in Figure 2-33). These leading edge folds add curvature to the wing, smoothing out the flight. These folds should not be creased. As you can see in the figures, they are more like bends than creased folds.

FIGURE 2-30   The plane with the tail section removed

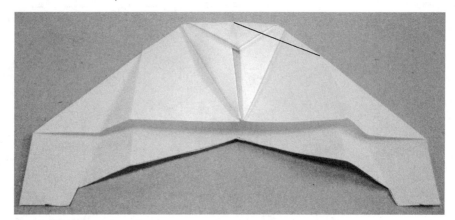

FIGURE 2-31   Place the ruler from the front centerline to the vertical wing fold to create the leading edge fold.

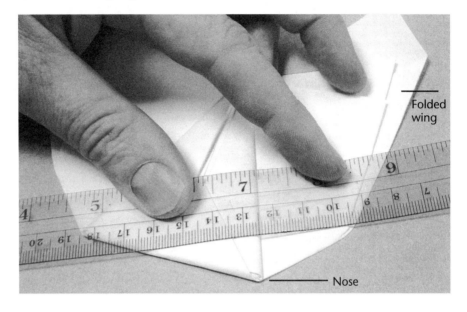

FIGURE 2-32   Use the ruler to fold the leading edges.

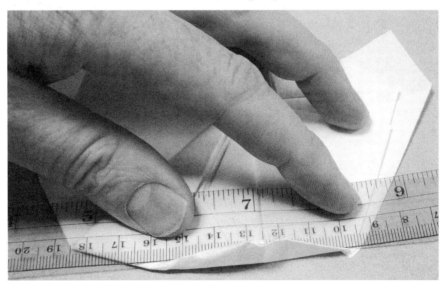

FIGURE 2-33   The top side of the airplane with leading edge folds

23. Fold the plane along the centerline, as shown in Figure 2-34.

24. Pull the wings outward to create a shallow angle to the centerline fold; the plane should look like Figure 2-35.

FIGURE 2-34 Fold the plane along the centerline.

FIGURE 2-35 The plane with completed centerline fold

25. Pinch the centerline to reverse the centerline fold from the trailing edge toward the midpoint along the centerline fold, as shown in Figure 2-36.

26. Place your thumb at the midpoint to make sure the centerline fold will remain upward, forward, and downward toward the trailing edge, as shown in Figure 2-37.

FIGURE 2-36   Reverse the centerline fold to the midpoint between the nose and the trailing edge.

FIGURE 2-37   Place your thumb at the midpoint on the centerline fold so the fold is upward, forward, and downward toward the rear of the plane.

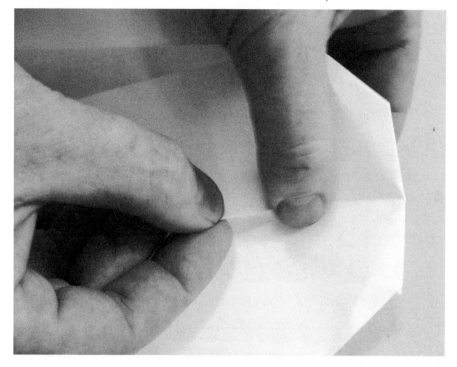

27. Refold along the centerline, making sure the two new folds resulting from the reverse fold are aligned, as shown in Figure 2-38. Once the two new folds are aligned, crease them down (Figure 2-39).

FIGURE 2-38 Refold along the centerline, making sure the two new folds are aligned.

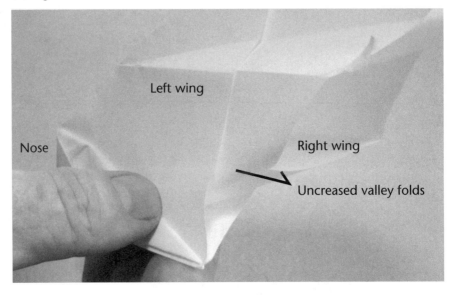

FIGURE 2-39 Crease down the folds to complete the speed brake fold.

28. Gently pull the wings outward once again to create a shallow angle to the centerline fold so the plane looks like Figure 2-40.

FIGURE 2-40 The completed speed brake fold

This completes the basic assembly of the surfer paper airplane. Next up is how to trim the surfer to fly straight and glide smoothly and slowly.

Trim the Paper Airplane Surfer

The tumblewing design in Chapter 1 requires little if any changes before it will fly properly. The paper airplane surfer will probably not fly well the first time, however, so it will need to be trimmed (adjusting the folds and elevons) to improve its flight. First we'll do a kind of course trim, making sure all the folds are more or less at the same angle from one side of the plane to the other. Next we'll use the trailing edge elevon tabs to fine tune the trim for maximum glide distance and trimming the plane to fly straight. Proper trim is a matter of life and death for flying animals and comes naturally to pilots of big airplanes. Trimming our model planes will enable us to fly them as walkalong gliders. To fly a walkalong glider, the glider will need to fly slow enough for us to keep up with it and smooth enough for us to predict where it will fly next to position the paddle in the right place.

Sight down the paper airplane surfer parallel to the centerline, as shown in Figure 2-41, noting any differences between the right and left leading edge folds, the vertical fins, and trailing edge trim tabs (elevons).

For example, Figure 2-42 shows the right vertical fin oriented more vertically than the left fin. This might cause the plane to turn toward the left. After you

FIGURE 2-41   Sight down the surfer from the front to check that each side is the same.

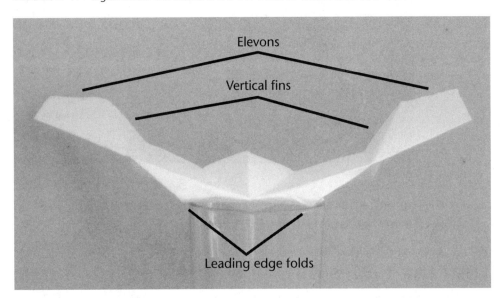

FIGURE 2-42   The right fin is more vertically oriented than the left fin, which might cause the plane to turn to the left.

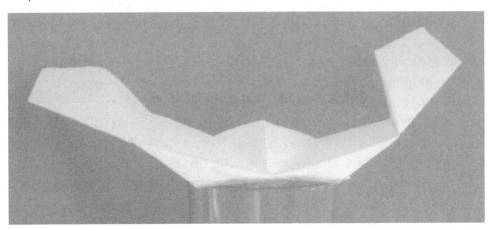

correct any differences between the right and left sides of the plane, launch the plane at a little faster than walking speed and with the nose pointing a little down. Notice whether the plane turns right or left, dives, or has a wavy flight path. The object of trimming is to make the plane fly as slowly and smoothly as possible with minimal right or left turn, diving, or wavering.

Use the ruler to draw a straight line across the tail, as shown in Figure 2-43.

In Figure 2-44, the drawn lines mark the elevon folds. You can fold along the lines to adjust the elevon tabs (see Figure 2-44).

FIGURE 2-43   Draw a line across the tail to mark the position of the elevon flap folds. An elevon combines the functions of elevators and ailerons on tailless aircraft designs.

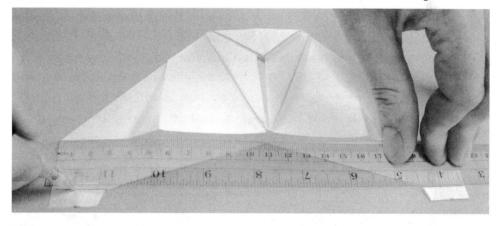

FIGURE 2-44   Lines showing elevon folds

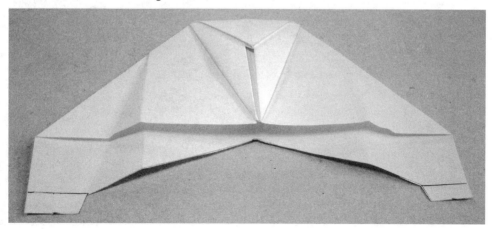

<u>NOTE</u>   *When adjusted in the same direction, the elevons function like an elevator control. The elevator control changes the angle the wings make with the oncoming air (this angle rotates about the pitch axis, discussed in the next section). To make the glider turn more to the right, raise the right elevon while lowering the left elevon. When adjusted in opposite directions, the elevons function like ailerons. The aileron control changes the bank angle of the wings (this angle rotates about the roll axis, which will be discussed in the next section).*

Trimming the Paper Airplane Surfer about the Pitch Axis

What's the pitch axis? Imagine your head is the airplane. The pitch axis would go horizontally through your ears and out over your shoulders. Rotating your head about this axis is like you are nodding your head when you say "yes." Figure 2-45 shows how the elevons need to be both moved up to pitch the nose up, in the same motion as when you start to nod your head. Figure 2-46 shows how the elevons both need to be lowered to pitch

FIGURE 2-45   Moving both elevons in the same direction affects rotation about the pitch axis. Moving both elevons up makes the nose pitch up.

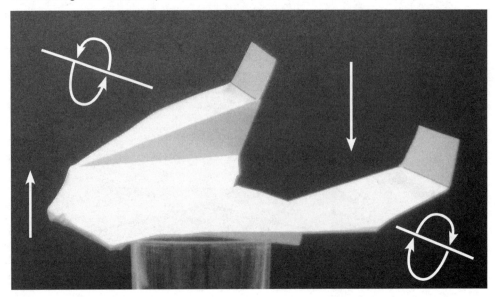

FIGURE 2-46   Moving both elevons down pitches the nose down.

the nose down in the same motion as when you finish nodding your head (your nose goes down too).

Thus the paper airplane surfer is trimmed (or adjusted) about the pitch axis by deflecting the elevons in the same direction (both up or both down as in Figures 2-45 and 2-46, respectively).

The pitch axis adjustment is important because it balances the four forces acting on an airplane—the forces of thrust and drag which act along the line of flight, and lift and weight which act up and down, respectively (see Figure 2-47). Perform another glide, launching your paper airplane a little nose down and trying to let go when the plane is at about walking speed. Figure 2-48 shows a side view of three test glides. If the plane dives, the elevons need to be adjusted a little upward. If the plane has a wavy flight path, the elevons need to be adjusted a little downward. When the plane is properly trimmed in the pitch axis it will have a smooth flight path and will glide the farthest distance.

FIGURE 2-47   There are four forces that need to be balanced for the airplane to fly smoothly without diving or having a wavy, rollercoaster-like flight path. Thrust and drag act along the direction of flight, whereas lift and weight act up and down.

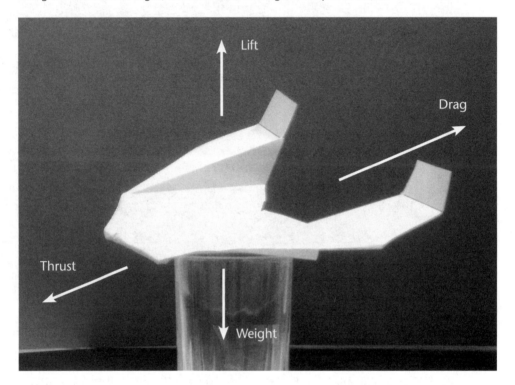

FIGURE 2-48   Side view of test glides showing smooth, wavy, and diving flight paths. The elevons are adjusted down to make the wavy flight path smooth or up to make the dive more shallow.

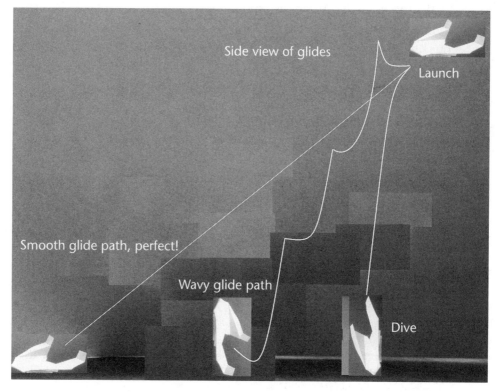

Trimming the Paper Airplane Surfer about the Roll Axis

What's the roll axis? Again, imagine your head is the airplane. The roll axis would go horizontally through your nose and out through the back of your head (Figure 2-49). Rotating your head about this axis is rocking your head side to side, shoulder to shoulder. This same head wobble means "I understand" in India.

The paper airplane surfer is trimmed (or adjusted) about the roll axis by deflecting the elevons in the opposite direction. Raising the right elevon and lowering the left elevon results in a turn to the right as shown in Figure 2-49. Raising the left elevon and lowering the right elevon results in a turn in the opposite direction, to the left.

Each flight has a problem either to the left or to the right and needs to be trimmed about the roll axis. If the flight path is to the left and smooth, the left elevon is adjusted a little down and the right elevon a little up. If the flight path is left and wavy (roller coaster) just move the left elevon down

FIGURE 2-49 Moving the elevons in the opposite direction will cause the plane to roll into a bank. Moving the right elevon up and the left elevon down will result in a roll to the right, with the left wing going up and the right wing going down. The roll axis of rotation is parallel to the direction of flight.

FIGURE 2-50   Back view of test glides of the paper airplane surfer.

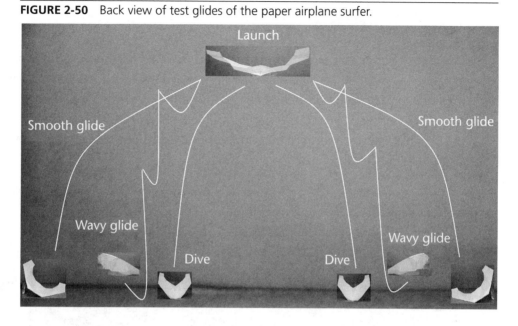

a little. If the flight path is left and diving, adjust only the right elevon up a little. For glides turning right the process is the opposite. See Table 2-1 for adjustments about both the roll and pitch axes.

Figure 2-50 shows six glides, three to the left and three to the right, of smooth, wavy, and dive flight paths. See Table 2-1 for what to do in each case. Note that adjustments can be made about two axes at once, pitch and roll, by moving just one of the elevons.

TABLE 2-1   Which Way to Trim the Elevons

		Roll Axis		
	Flight Path	**Left Turn**	**Straight, yes!**	**Right Turn**
Pitch Axis	**Wavy flight**	Left down	Left and right down	Right down
	Smooth, yes!	Left down, right up	Perfect!	Left up, right down
	Dive	Right up	Right and left up	Left up

The optimum gliding speed is as slow as possible, while maintaining a smooth glide.

Once your paper airplane surfer is trimmed to fly straight, smoothly, and as slowly as possible (as shown in the flight path in Figure 2-51), you are ready to try sustaining and controlling your airplane with the paddle you made in Chapter 1. Flying the paper airplane surfer using the paddle makes this a controllable paper airplane and your ticket to becoming a true paper airplane pilot!

FIGURE 2-51   A properly trimmed paper airplane surfer will glide smoothly in a relatively straight line.

Keep Your Paper Airplane Surfer in the Air

The paper airplane surfer is sustained in the air in a similar way as the tumblewing described in Chapter 1. The paddle (a large sheet of cardboard) is used to generate a wave of rising air as you, the pilot, follow the paper airplane surfer as it flies. By keeping the paper airplane surfer close to the top edge of the paddle, you can keep your paper airplane surfer in the air for a long time (see Figure 2-52).

Without the paddle, the paper airplane surfer will glide diagonally toward the ground, as shown in the flight trajectory of Figure 2-53.

FIGURE 2-52   Keep your paper airplane surfer in the air by orienting the top edge of the paddle close to the flying plane.

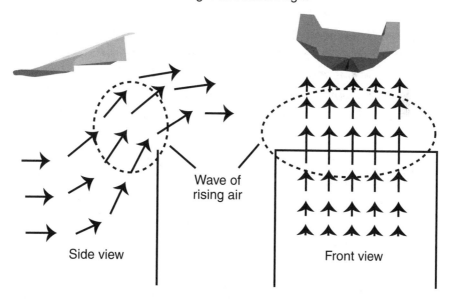

Straight and Level Flight

Wave of rising air

Side view

Front view

FIGURE 2-53 Paper airplane surfer glide path without paddle

Launch the paper airplane surfer from as high as your arm can reach, and aim for a speed as close to the plane's gliding speed (at about walking speed) as possible. It also helps to point the nose of the plane down a bit as you let it go. As soon as you have launched the plane, start moving the paddle at about your eye level. Move smoothly, following along with the plane, and keeping the paddle's top edge close to the tail of the plane. At this point, the wave created by the paddle should be keeping the plane in the air, as shown in the flight trajectory of Figure 2-54.

FIGURE 2-54 The paper airplane surfer flight path is sustained by the wave created by the top edge of the paddle.

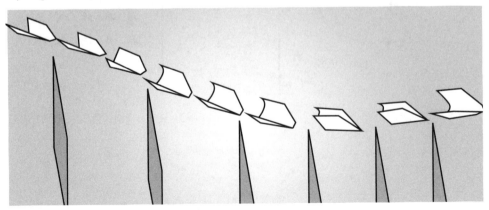

Steer Your Paper Airplane Surfer

To turn your paper airplane surfer, tilt the paddle in the direction of the turn (see Figure 2-55). This creates more lift under one wing, which results in the plane turning toward the lower wing.

The plane may not react quickly to the tilted paddle, so have patience and keep trying. By alternating the direction of the paddle's tilt, you can turn the airplane in both directions, as shown in the flight trajectory of Figure 2-56.

<u>NOTE</u> *If the paper airplane surfer naturally turns to one direction, you can compensate by moving the paddle, but it will be more difficult to make it turn in the opposite direction.*

FIGURE 2-55 To turn the paper airplane surfer, tilt the paddle to produce more lift under a wing.

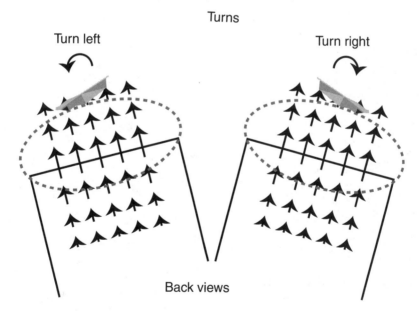

FIGURE 2-56 Change the paddle's tilt to alter the flight path of the paper airplane surfer; here the paddle's tilt causes the plane to make several S-turns.

Make Your Paper Airplane Surfer Climb

You can make your paper airplane surfer climb by placing the paddle closer to the tail of the plane to maximize the rising air (see Figure 2-57)—just make sure you don't crash the paddle into the plane.

FIGURE 2-57   Move the paddle as close as possible to the tail of the plane.

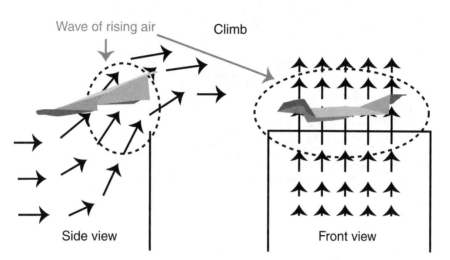

Make a Paddle Launcher

You can make a simple paddle launcher to launch your paper airplane surfer using two jumbo paper clips. The launcher is mounted on the top edge of the paddle (see Figure 2-58), with the airplane surfer resting on these clips.

After you mount the launcher and begin moving the paddle, as the paddle gathers speed, the paper airplane surfer takes off into the rising air of the paddle. The nose of the plane is held up so it will fly directly into the rising air created by the paddle as it gathers speed.

What you'll need (see Figure 2-59):

- Two jumbo paper clips (5 cm, or 4 in., long)

- Cellophane or masking tape

- Ruler

FIGURE 2-58   The paper airplane surfer on the paddle launcher

FIGURE 2-59   What you will need to make a paddle launcher

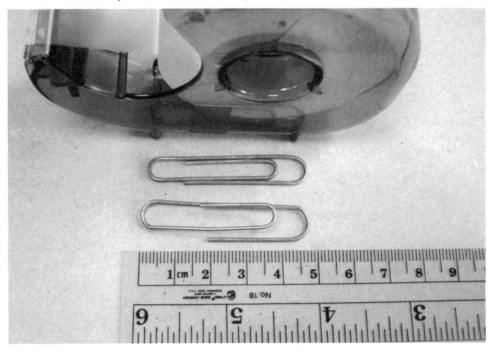

Assembly

1. Separate the large and small loops of a jumbo paper clip, as shown in Figure 2-60.

2. Bend the small loop until it is at a right angle (90 degrees) from the large loop, as shown in Figure 2-61. Do the same to the other jumbo paper clip; both clips are shown in Figure 2-62.

FIGURE 2-60   Separate the large and small loops of the jumbo paper clips with your finger.

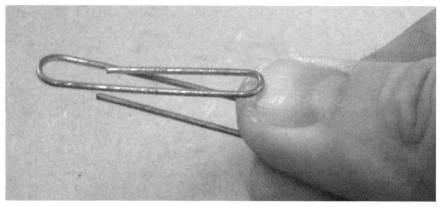

FIGURE 2-61   Bend the small loop to a right angle relative to the large loop of the jumbo paper clip.

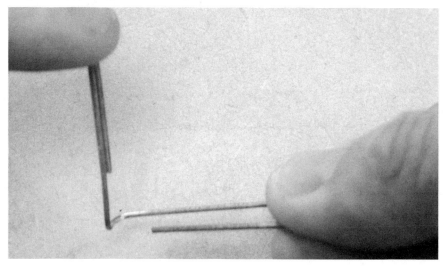

FIGURE 2-62   Both paper clips

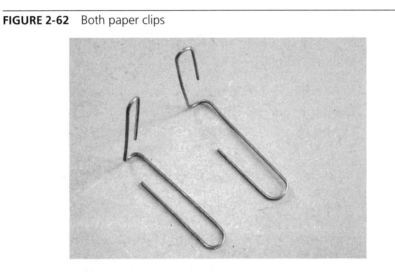

3. Tape down both paper clips about 6 cm (4 1/2 in.) apart at the midpoint of the top edge of the cardboard paddle, as shown in Figure 2-63.

FIGURE 2-63   Tape the large loops of the paper clips to the midpoint of the top edge of the cardboard paddle.

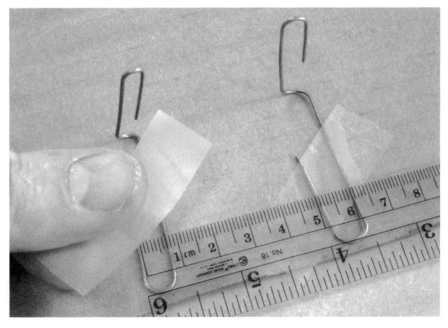

FIGURE 2-64   The completed paddle launcher.

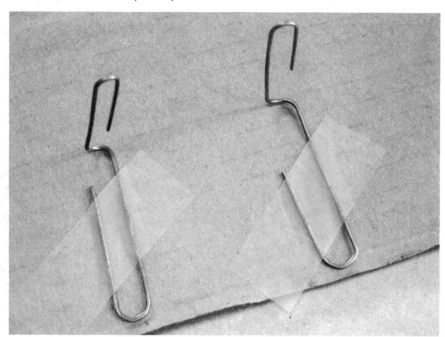

That's all you need to do!

Figure 2-64 shows the completed paddle launcher.

Launch the Airplane from the Paddle Launcher

It can be a little difficult to launch the paper airplane surfer from the launcher, because the plane has little time and altitude to stabilize. Keep practicing!

1. Hold the paddle so the paper airplane surfer on the launcher is horizontal, as shown in Figure 2-65.

2. Tilt the paddle forward so the airplane is pointing nose down (but not so much that the plane slides off the launcher).

3. Start moving, and as soon as the plane lifts off, tilt the paddle to the vertical with the top edge close to the tail of the paper airplane surfer, as shown in Figure 2-66.

FIGURE 2-65   Position of the paddle so the paper airplane is propped up by the paddle launcher.

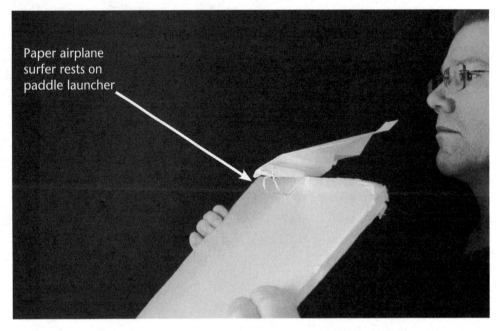

Paper airplane surfer rests on paddle launcher

FIGURE 2-66   After the airplane is launched, tilt the paddle to the vertical position.

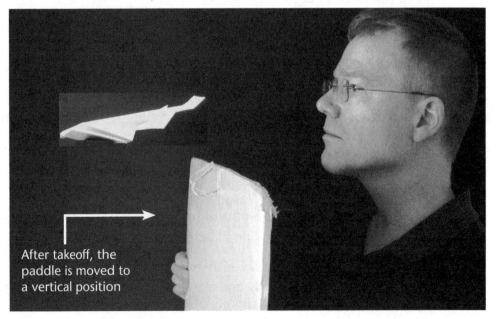

After takeoff, the paddle is moved to a vertical position

FIGURE 2-67   Flight path of the airplane after launch

Figure 2-67 shows the flight path of the paper airplane surfer being launched from the paddle.

NOTE   In the next chapter, the X-surfer foam glider can also be launched from the paddle launcher.

Theory

The paper airplane surfer relies on its forward motion through the air to sustain its flight. As the forward speed slows, the angle the wing makes into the oncoming air must be increased (Figure 2-68).

Notice that the wing tips are at a lower angle to the oncoming air than the central wing section. There are two reasons for this: First, when the forward motion of the plane slows to the point where the air no longer supports the plane, because of its increased angle, the central section will stop producing lift before the wing tips do, resulting in a straight and forward path as the plane drops and gathers speed again. The second reason for the tip angles is to facilitate turning the paper airplane surfer.

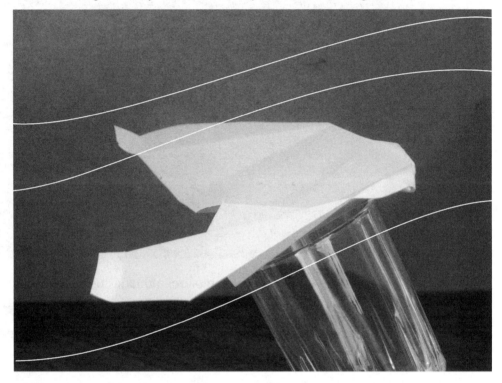

Figure 2-69 shows a paper airplane surfer with the paddle tilted to roll the plane into a turn. To roll the paper airplane surfer into a turn, you tilt the paddle in the direction of the turn. The closer the paddle comes to the wing, the stronger the upward moving air, and the greater the angle the wing makes with the oncoming air. In a turn, one wing will experience a higher angle relative to the oncoming air than the other wing.

The wing tips must continue to produce lift, especially the tip closest to the paddle. Having a reduced angle at the wing tip keeps the wing tip flying and prevents the tip from suddenly dropping because of the increased angle with the oncoming air.

FIGURE 2-69   The arrows show the increased upward motion of the air close to the top edge of the paddle.

Experiment 1: Paper Grain Direction

Certain types of paper are stronger in one orientation than in another. To see this in action, try this experiment.

1. Cut a page from a phone book or other lightweight newsprint, and place the page lengthwise over the gap between two books, as shown in Figure 2-70.

2. Increase the gap between the books until the page just begins to sag. Now rotate the page 90 degrees so the shorter width is spanning the same gap. Does the shorter dimension hold up as well or does it sag a lot more, as shown in Figure 2-71? Phonebook paper has a grain oriented parallel to the spine of the phone book, as shown in Figure 2-72.

FIGURE 2-70   Place a piece of phonebook paper lengthwise over the gap between two books.

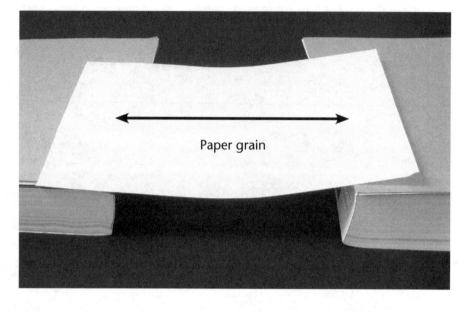

FIGURE 2-71   Rotate the page so the width is now bridging the gap; the page sags a lot more, because of the grain of the paper.

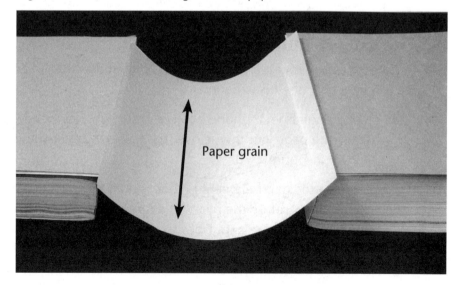

FIGURE 2-72   The grain of the paper is usually parallel to the spine of the book.

3. To make a stronger paper airplane, cut the page so the long side of the page is perpendicular to the long side of the page, as shown in Figures 2-73 and 2-74. This orients the paper grain perpendicular to the wingspan of the paper airplane surfer and lessens the tendency for the wings to fold up in flight. Orienting the rectangle this way also results in a smaller sized plane, and the paper is effectively more rigid.

FIGURE 2-73   Orient the long edge of the rectangle perpendicular to the grain of the paper to strengthen the wings and make them less likely to fold up in flight.

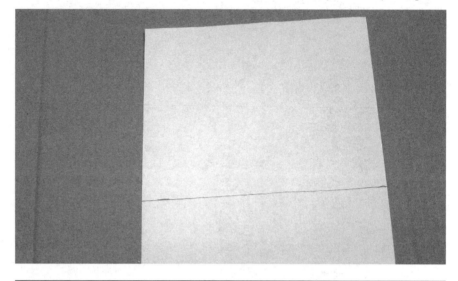

FIGURE 2-74   Cut the page so the long side of the rectangle is perpendicular to the grain of the paper.

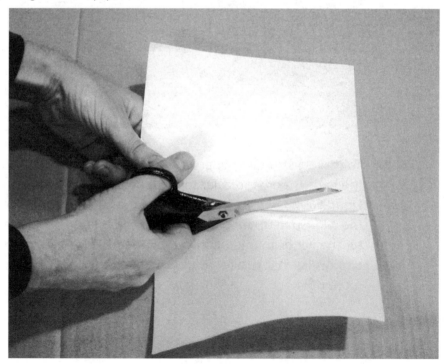

Experiment 2:
Wing Loading of Different Designs

What would happen if you made different paper airplane designs with the same weight paper?

Suppose you made a tumblewing (Chapter 1) and a paper airplane surfer out of the same weight paper. You know that the tumblewing has roughly the same area as the paper used to make it (minus the up-folded winglets). The paper airplane surfer has extra nose weight created from folding the sheet of paper over on itself multiple times. Each time the paper is folded, the area is reduced, but the mass remains the same. The paper airplane surfer will fall much faster than the tumblewing design made from the same weight paper, because of the airplane surfer's nose weight and increased wing loading.

3

X-Surfer

Project size: Medium   **Skill level:** ★★★☆

The X-surfer (see Figure 3-1) is a flying wing design invented by Tyler MacCready made from Depron foam. The foam used in the X-surfer is shaped using the heat from a household iron (this process is called *thermoforming*). The X-surfer's nose weight is made from a common paper clip.

What you'll need:

- Sheet of lightweight 1 mm thick Depron foam (at least 40-by-15 cm, or 15 3/4-by-6 in.; you can purchase the sheets at www.rcfoam.com or at hobby shops)

FIGURE 3-1   Completed X-surfer

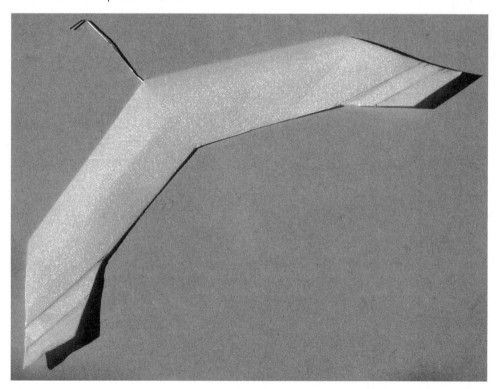

- Large sheet of cardboard (1 square meter, or 3 feet to each side [9 square feet], is optimal); the cardboard used in Chapters 1 and 2 can be reused

- Magazine or cardboard about 3 mm (1/8 in.) thick

- Ruler (marked in centimeters and/or inches)

- Scissors

- Pen

- Cellophane tape

- Household iron

- 3 cm (1 1/4-in.) paper clip (preferably steel and attracted to a magnet)

Assembly

1. Make copies of the patterns for the right and left wings, as shown in Figures 3-2 and 3-3, respectively. It is important that the dimensions of each printed pattern be about 15-by-20 cm (about 6-by-8 in.); this ensures that the size of the airplane will work with the paperclip nose weight.

    NOTE   *These patterns can also be downloaded from the web site at www.walkalongglider.info.*

    Each wing pattern should be about 19–20 cm (about 8 in.) wide, measuring from a line extending from the wing root to the wing tip, as shown in Figure 3-4.

2. Cut along the thick lines of the right and left wing patterns (see Figure 3-5). The end result should be two patterns for the right and left wings, as shown in Figure 3-6.

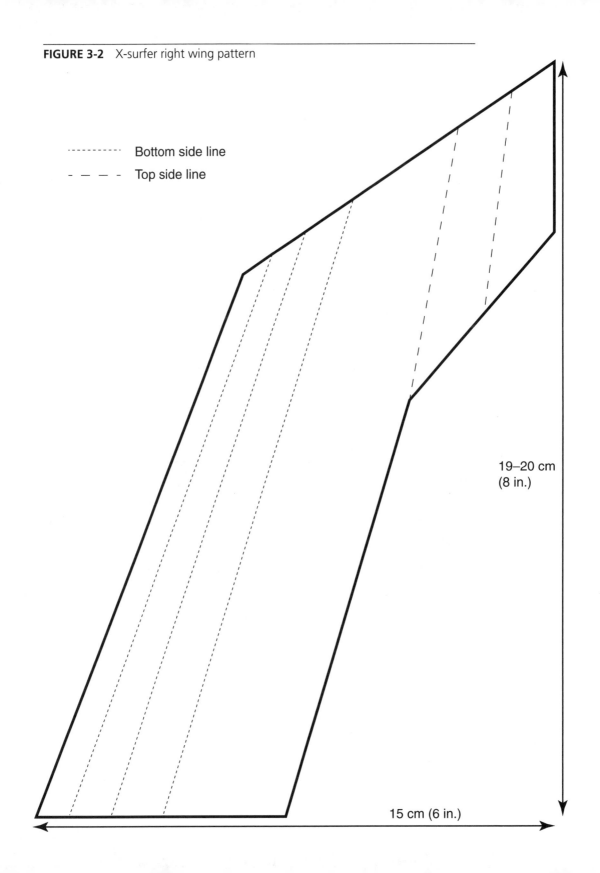

FIGURE 3-2 X-surfer right wing pattern

Bottom side line

Top side line

19–20 cm (8 in.)

15 cm (6 in.)

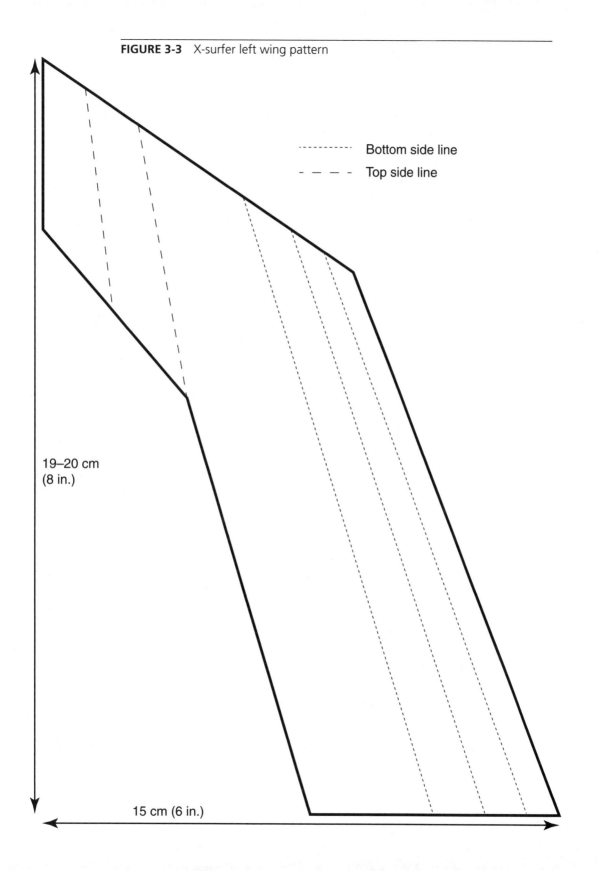

FIGURE 3-3 X-surfer left wing pattern

.... Bottom side line

– – – – Top side line

19–20 cm
(8 in.)

15 cm (6 in.)

FIGURE 3-4 The right wing pattern with a ruler for scale (here the pattern measures 19.3 cm); the final wingspan should be about 38–40 cm (15 3/4 in.) total.

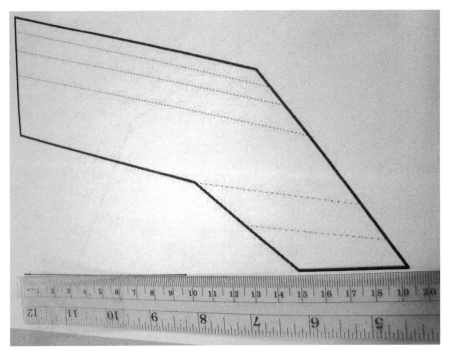

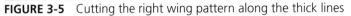

FIGURE 3-5 Cutting the right wing pattern along the thick lines

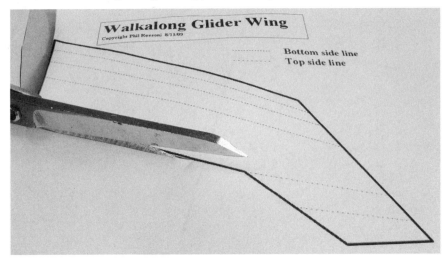

FIGURE 3-6   Wing patterns for the X-surfer wings

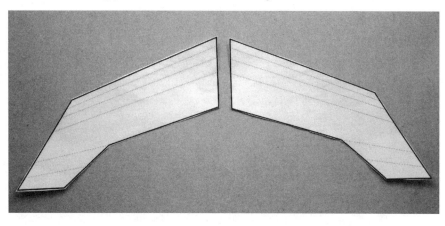

FIGURE 3-7   Tape together the right and left wing patterns.

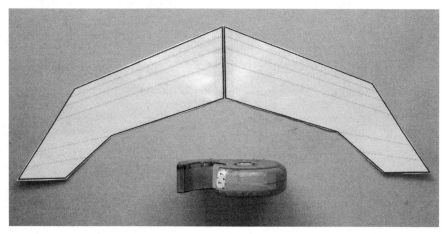

3.   Tape together the right and left wing patterns, as shown in Figure 3-7.

4.   Tape the pattern to the Depron sheet, as shown in Figure 3-8.

FIGURE 3-8 Tape the pattern to the Depron sheet.

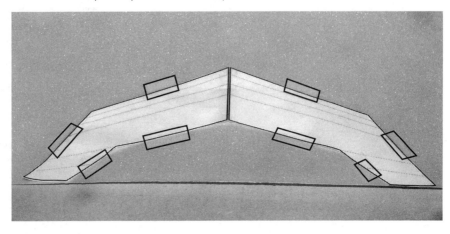

FIGURE 3-9 Cut the Depron sheet along the pattern's outline.

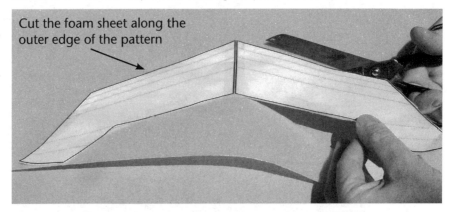

Cut the foam sheet along the
outer edge of the pattern

5. Cut out the Depron sheet along the outlines of the pattern, as
   shown in Figure 3-9. If necessary, add extra tape to make sure the
   paper pattern stays attached to the Depron sheet.

   Transfer the airfoil bend lines to the Depron sheet. This is the
   bottom side of the wings. Start by marking indentations where the
   three airfoil crease lines meet the centerline, as shown in Figure
   3-10. An airfoil is the curved shape of a wing which lifts the airplane
   by pushing down on the oncoming air.

6. Mark the end of each line at the right wing tip edge, as shown in
Figure 3-11. Do the same for the left wing tip edge.

FIGURE 3-10   Mark indentations along the centerline at the three airfoil
bend lines.

FIGURE 3-11   Mark the end of the three airfoil lines on the edge of the right
wing tip.

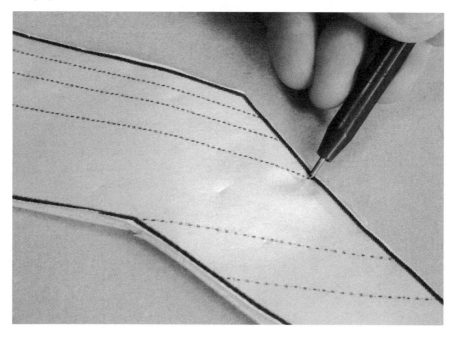

7. Remove the pattern and connect the dots on the Depron sheet to mark the three airfoil crease lines on the right and left wings, as shown in Figure 3-12. The Depron sheet wings should now look like Figure 3-13.

FIGURE 3-12   Mark the three airfoil bends on the surface of the Depron sheet.

FIGURE 3-13   The three airfoil bend lines on the bottom of the Depron sheet

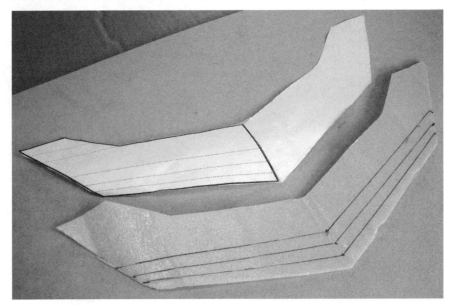

FIGURE 3-14   Use the pattern to transfer the elevon bend lines to the other side (the top) of the Depron sheet.

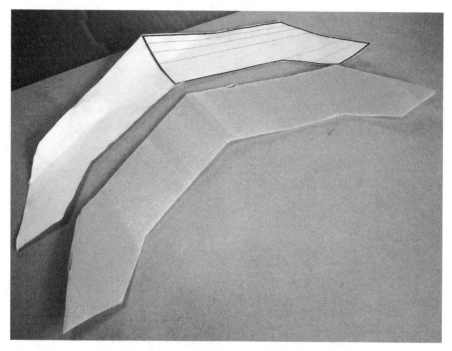

8. Turn over the Depron sheet. Now you'll use the pattern to mark the elevon bend lines on the top surface (see Figure 3-14) of the Depron sheet.

9. Mark the ends of the elevon lines on the right and left wing edges, as shown in Figure 3-15.

FIGURE 3-15   Mark the inboard end of the elevon line for the left wing.

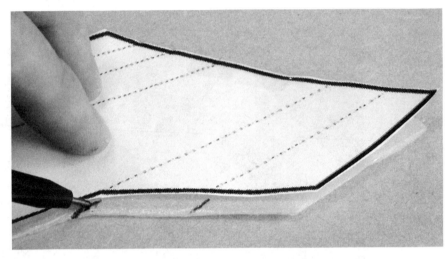

FIGURE 3-16   Mark the elevon line on the top surface of the Depron sheet.

10. Remove the pattern and connect the lines on the top surface, as shown in Figure 3-16, for both right and left elevons. The top surface of the Depron sheet should now look like Figure 3-17.

11. Preheat a household iron to a temperature suitable for silk and acrylic, as shown in Figure 3-18.

FIGURE 3-17   The elevon markings on the top surface of the Depron sheet

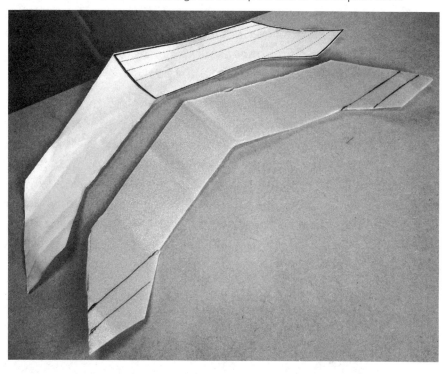

FIGURE 3-18   Preheat the iron.

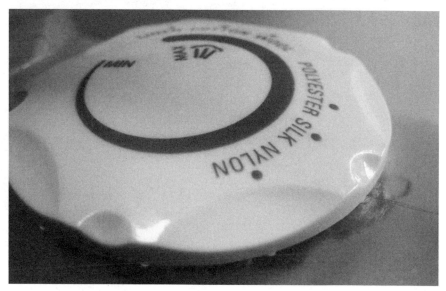

<u>*CAUTION*</u>   *The iron will be hot, so avoid contacting the surface. Adult supervision is recommended. Also, work in a well-ventilated area, because thermoforming polystyrene produces harmful vapors. Also use a firm smooth surface or second magazine to iron on to protect the table top from the heat.*

Now you'll need a straight edge about 3 mm (1/8 in.) thick, such as the edge of a magazine or a piece of cardboard. You'll use this edge to thermoform each airfoil bend. Practice thermoforming a slight bend in a piece of scrap Depron. If the iron sticks to the Depron, the iron is too hot. If the Depron does not give easily, the iron is not hot enough. Align the first line on the foam wing parallel to but 3 mm (1/8 in.) away from the straight edge, as shown in Figure 3-19. Once again, the angle of the bend is slight as the airfoil is made up of three such bends.

FIGURE 3-19   Use a straight edge (such as a magazine) to thermoform the airfoil bends in the Depron sheet.

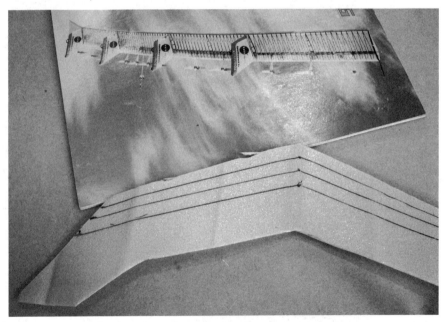

Start the bend by placing the hot iron where the line meets the wing centerline, as shown in Figure 3-20. Do not press hard; you may have to wait for the iron to heat the Depron.

Move the iron slowly along the line, making sure that the line stays 3 mm (1/8 in.) from the straight edge. Note where the straight edge emerges on both ends and keep this position as the iron moves down the wing tip. A thermoformed bend will be created in the gap between the straight edge and the work surface, as shown in the side view in Figure 3-21. The first thermoformed airfoil should look like Figure 3-22.

FIGURE 3-20   Start each airfoil bend where the line meets the centerline. The iron will bend the Depron. You will feel the depron get weaker and give way as it heats under the iron.

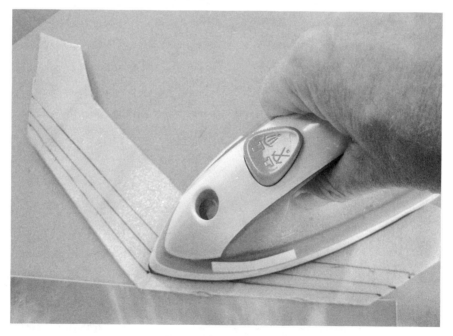

FIGURE 3-21 Side view of the iron thermoforming a bend in the Depron sheet in the gap between the straight edge and the work surface

FIGURE 3-22 The Depron sheet after thermoforming the first airfoil bend

Thermoform another airfoil bend on the other wing, also starting at the centerline, as shown in Figure 3-23.

12. Continue with the second airfoil crease line in the same way. Both creases in the Depron sheet should look like Figure 3-24.

FIGURE 3-23   Create another airfoil bend on the other wing.

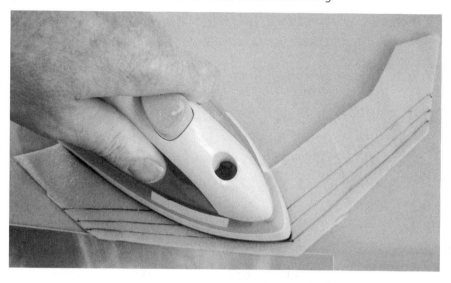

FIGURE 3-24   The Depron sheet after finishing both airfoil bends

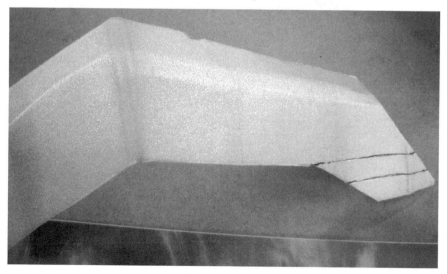

13. Continue with the rest of the airfoil crease lines so the top surface of the Depron sheet looks like Figure 3-25.

14. Next turn over the Depron so the elevon lines show and thermoform the two elevon creases for each wing in the same way as the airfoil creases, as shown in Figure 3-26.

FIGURE 3-25   The top surface of the Depron sheet after finishing all three airfoil bends

15. As discussed in the theory section of Chapter 2, the wing tips need a lower angle to the oncoming air as compared to the wing root (or centerline). Whereas the paper airplane surfer had a sudden change in the angle of the wing at the vertical fin, the X-surfer will have a gradual twist along the length of the wing (see Figure 3-27). Use

FIGURE 3-26   Thermoforming the first elevon bend of the left wing

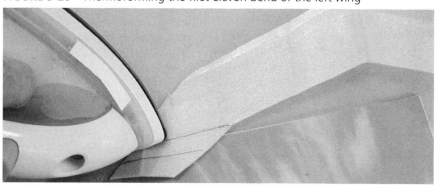

FIGURE 3-27   Comparison of paper airplane surfer and X-surfer wing washout. The paper airplane surfer has a step change in wing twist at the vertical fin. The X-surfer has a gradual twist along the length of the wing.

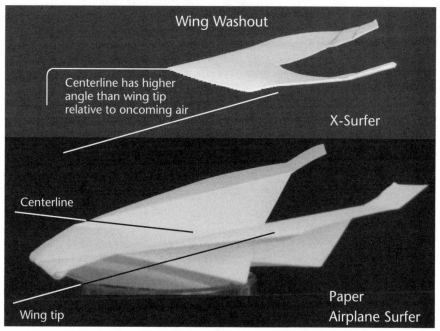

the iron to thermoform a wing washout angle to the left and right wings, as shown in Figures 3-28 and 3-29. To do this, twist the wing nose up, trailing edge down as you draw the wing out from under the hot iron.

FIGURE 3-28 Thermoforming the wing washout angle to the left wing. Twist the nose up and trailing edge of the wing down while drawing the wing out from underneath the iron.

FIGURE 3-29 Thermoforming the wing washout angle to the right wing

16. Angle the iron so its tip touches the centerline of the nose of the X-surfer (the pointed top end). The back of the iron should not touch the foam's surface. Pull the nose of the Depron sheet up so the top surface of the sheet is touching the bottom of the iron. Be careful not to burn yourself!

17. Keeping the nose bent up, draw the polystyrene sheet out to thermoform an upward twist to the wing. This twist, called the washout angle, gives the centerline a slightly larger upward-facing angle than the wing tips with respect to the oncoming air in flight. This design feature helps the wings stay level as the glider slows down and will allow you to use the paddle to roll the glider into a turn.

18. Next create a nose weight using a standard paper clip (about 3 cm, or 1 1/4 in., in length), as shown in Figure 3-30. Note that if you use an aluminum paper clip you may need to add extra weight to the tip to compensate for the difference in density between steel and aluminum. Also, aluminum paper clips will not be attracted to a magnet. Uncoil the paper clip, as shown in Figure 3-31.

FIGURE3-30   The nose weight is made from a standard paper clip.

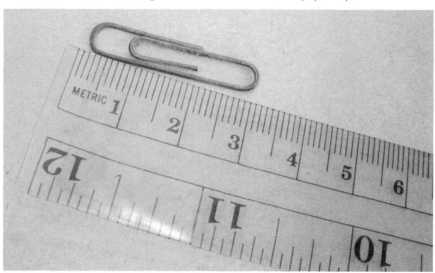

19. Leave a right angle bend at the end of the clip, as shown in Figure 3-32, so the end will be less likely to puncture or stab anything the glider runs into.

FIGURE 3-31   Uncoil the paper clip to make the nose weight for the nose of the X-surfer glider.

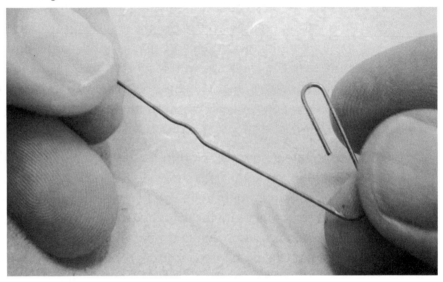

FIGURE 3-32   The bend is a right angle so the paper clip will be less likely to puncture whatever the glider runs into. A glider with a point on the end can hurt if it hits you.

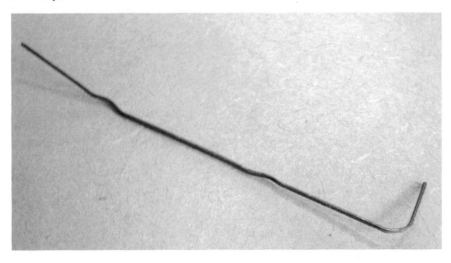

FIGURE 3-33   Attach the paper clip weight to the nose of the glider using transparent tape.

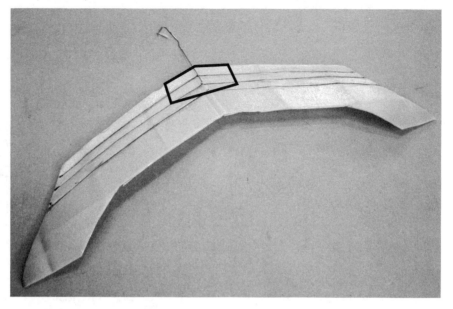

20. Tape the paper clip weight to the nose of the glider, as shown in Figure 3-33.

21. Slide the nose weight as far forward as possible. The X-surfer glider is now ready for gliding tests to adjust the weight position and elevons.

Trim Your X-Surfer

As with the paper airplane surfer (Chapter 2), the X-surfer needs to be trimmed to adjust the elevons and nose weight so the glider flies smoothly, slowly, and straight. As a walkalong glider, the X-surfer will need to fly on its own indefinitely in the updraft you create with the paddle, and, as such, it needs to be balanced. It is easier to move the ballast forward or back to get the X-surfer flying slowly and smoothly before trimming the elevons.

As was done with the paper airplane surfer, before performing a test glide, sight down along the nose of the X-surfer and note any asymmetries between the right and left elevons and wings, as shown in Figure 3-34.

FIGURE 3-34 View down along the nose of the X-surfer to check for asymmetries between the right and left wings.

See if you can predict which way the X-surfer will fly just by looking at it from the point of view of the oncoming air. Perform a test glide, launch the glider at about walking speed with the nose pointing slightly down. If the X-surfer dives, move the nose weight back so that less of it is sticking out in front. If the glider slows and then dives, as shown in Figure 3-35, adjust the nose weight forward so that more of it is sticking out in front. If the nose weight is as far forward as possible and the glider still slows and then dives, try thermoforming both elevons a little flatter. Figure 3-36 shows a reasonable glide trajectory.

The elevons of the X-surfer are trimmed in the same way as the paper airplane surfer (see Table 2-1). To correct turns the elevons are moved in opposite directions, as shown by the arrows in Figure 3-37. The X-surfer's speed can be controlled by raising or lowering the elevons in tandem, as shown by the arrows in Figure 3-38, although it is easier to move the nose weight forward or backward if possible.

FIGURE 3-35   If the X-surfer is trimmed too nose high, the nose will drop suddenly when the glider slows too much and can no longer fly. This situation can be corrected by moving the boom ballast forward and/or lowering the elevons.

FIGURE 3-36   A test glide of the X-surfer showing a slight turn to the right

FIGURE 3-37   The elevons of the X-surfer function as ailerons when they are moved in two different directions. Raising the left elevon relative to the right will make the glider turn more to the left.

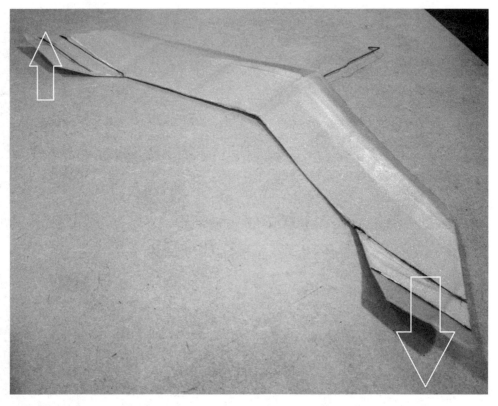

If the glider nose dives or flies too fast, you can raise the elevons and/or move the boom ballast back.

Sustain Your X-Surfer in Flight

As with the paper airplane surfer, the X-surfer flies relatively fast, and you will need a large open space with few drafts, such as a gym, in which to learn to fly. Once your X-surfer is trimmed to fly slowly, smoothly, and straight, you are ready to try sustaining it in the air using the paddle you made in Chapter 1. As with the tumblewing and paper airplane surfer gliders, the X-surfer glides downward without the paddle, as shown in Figure 3-39.

FIGURE 3-38 Raising both elevons of the X-surfer will raise the nose of the glider and result in a slower gliding speed. If the elevons are raised too much, the glide speed will be too slow and the nose will drop suddenly.

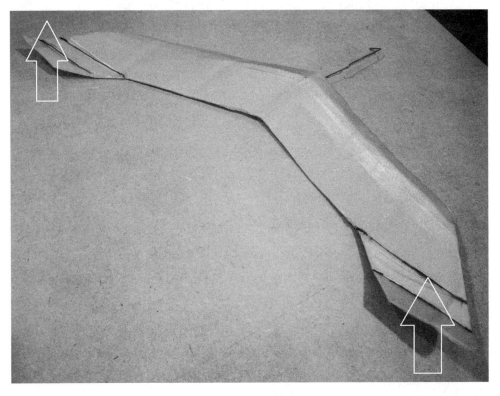

FIGURE 3-39 Flight path of the X-surfer without the wave of the paddle to sustain it

Keeping the paddle at eye level, launch the X-surfer from as high as you can at a fast walk speed , and angle the nose slightly down. Start moving behind the X-surfer as it descends, placing the top edge of the paddle as close to the flying X-surfer as you can in a steady manner, as shown in Figure 3-40.

The point of placing the paddle below and behind the trailing edge is to get the flying X-surfer in the area of maximum upward-moving air created by the paddle, as shown in Figure 3-41. Move the paddle as smoothly as possible. If you wave the paddle, the updraft will be unsteady and you are more likely to lose control of the glider.

FIGURE 3-40   X-surfer flight trajectory being sustained by the paddle following close behind and below

FIGURE 3-41   Diagram of front and side views of X-surfer riding the wave of air at the top edge of the paddle

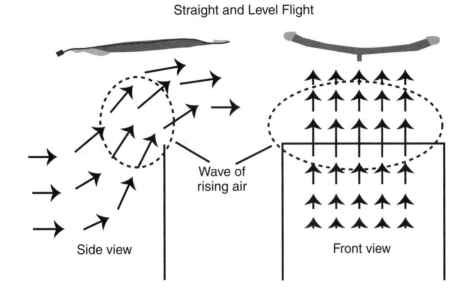

Straight and Level Flight

Wave of rising air

Side view

Front view

Steer Your X-Surfer

The X-surfer is turned in much the same way as the paper airplane surfer, by tilting the paddle in the direction of the turn, as shown in Figure 3-42.

Tilting the paddle right or left results in a turn to the right or left, respectively. The X-surfer may take time to react, so keep the paddle tilted in the direction until the X-surfer is clearly in the turn. As the X-surfer turns, anticipate the turn back to straight-and-level flight by applying opposite tilt before your desired heading is reached. Once a turn is completed in one direction, try tilting in the opposite direction to produce an S-turn, as shown in Figure 3-43.

FIGURE 3-42 Diagram showing how the paddle placement affects turns

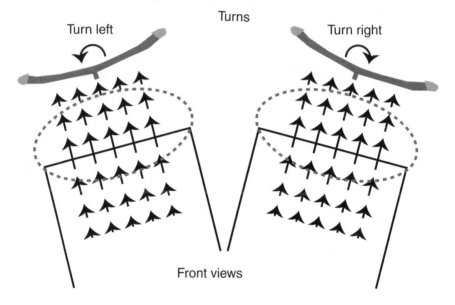

FIGURE 3-43 X-surfer S-turn flight trajectory

Make Your X-Surfer Climb

You will have more control over the X-surfer if you keep the glider flying high. As with the paper airplane surfer, you can make your X-surfer climb by placing the top edge of the paddle as close to the tail of the glider as possible without making contact, as shown in Figure 3-44.

Figure 3-45 shows the X-surfer in a climbing flight trajectory. Notice that the nose of the X-surfer in a climb does not point up. The glider always descends relative to the air current; the nose points down, so the air behind the glider is rising at a faster speed than the X-surfer is gliding down.

FIGURE 3-44   To make the X-surfer climb, place the top edge of the paddle as close as possible to the tail of the X-surfer.

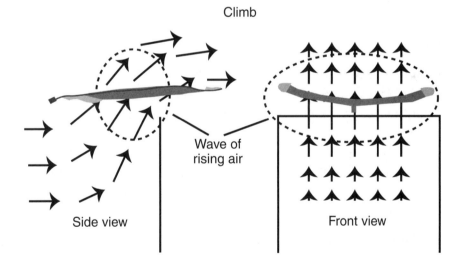

Climb

Wave of rising air

Side view

Front view

FIGURE 3-45   Flight trajectory of a climbing X-surfer

Theory

All airplanes need to be balanced, or trimmed, to fly properly. If you are preparing for a flight on a small plane, the flight attendant might ask for volunteers to move either to the front or to the back of the plane. This is necessary to balance the weight over the wings and functions in the same way as the X-surfer's movable ballast—the paper clip, which can be slid forward and back relative to the wings. The center of lift of the wings, where all the lift forces balance, is approximately on a line that divides the wing area in two, as shown in Figure 3-46.

If dropped in the horizontal position without the nose weight, the wing would probably alternate between dropping its tail and nose, like a falling leaf or feather (it may even tumble in much the same way as a tumblewing). With the nose weight in place, the nose will drop first. As the wing gathers speed, it starts to fly forward, and the elevons produce a downward push on the tail. This push becomes larger as the glider speeds up, balancing the downward force of gravity on the nose weight. So the aerodynamic force of the elevons balances the constant force of the nose

FIGURE 3-46   In this diagram, the center of lift is the point at which the vertical centerline intersects with a horizontal line, where half the wing area is in front and half is behind the horizontal line.

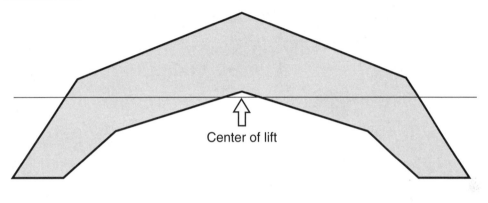

Center of lift

FIGURE 3-47   X-surfer with nose weight adjusted fully back. In this configuration, the X-surfer's center of weight (or balance point) is closer to the tail.

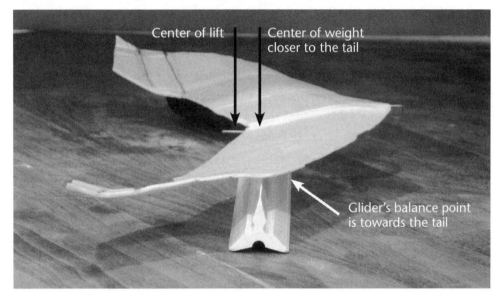

Center of lift   Center of weight
                        closer to the tail

Glider's balance point
is towards the tail

weight, with the nose rising as speed increases and dropping as the glider slows. Moving the nose weight back slows the speed of the X-surfer as the balance favors the elevons' ability to raise the nose (Figure 3-47).

Moving the nose weight forward (Figure 3-48) results in a faster speed before the elevons balance the glider.

FIGURE 3-48 X-surfer with nose weight as far forward as possible. In this configuration the X-surfer's center of weight (or balance point) is toward the nose and is more stable.

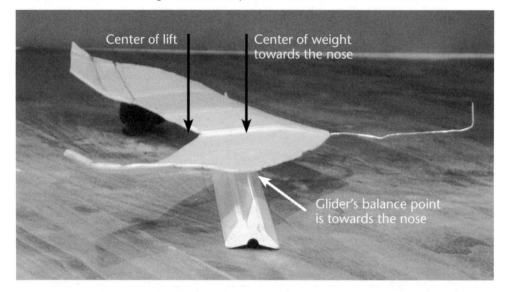

To trim the nose weight, start with the weight as far forward as possible and adjust it back until the glider can glide smoothly at a slow speed. This speed will also be approximately the best glide speed so the X-surfer will travel the farthest distance. Why do we use an uncoiled paper clip instead of a weight that doesn't stick out so far? Balancing the glider by placing the weight as far forward as possible allows the X-surfer to be balanced with less weight so it can fly slower. In Chapter 4, we'll build the jumbo, which has a wingspan of more than 1 meter (3 ft.)!

4

Jumbo

Project size: Large   **Skill level:** ★★★★

Here's the big daddy of them all: a 1.2-meter (47 in.) wingspan jumbo glider, shown in Figure 4-1. Flying silently, the jumbo has the potential to frighten unsuspecting people, awakening self-preservation instincts and sending them running for cover wondering, What if that big flying animal is hungry? When flying the jumbo, you will experience handling a larger and heavier airplane that does not respond as quickly as smaller gliders. Because of its size, the jumbo is heavier and will fly correspondingly faster. The difference between flying the tumblewing and the jumbo, for example, is a bit like the differences in flying a light trainer such as a Cessna 150 and an airliner such as a Boeing 747 or A-380. It really does take an area the size of a gym to turn around the jumbo.

What you'll need:

- 70-by-49.5 cm (27 1/2-by-19 1/2 in.) sheet of 3 mm thick Depron foam, available at hobby stores or at www.rcfoam.com (see Figure 4-2)

FIGURE 4-1   Completed jumbo walkalong glider

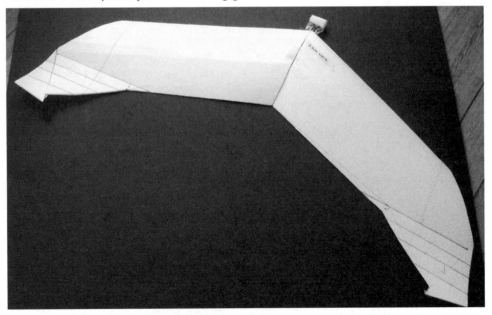

FIGURE 4-2   The 70-by-49.5 cm (27 1/2-by-19 1/2 in.) sheet of 3 mm thick Depron foam from which the jumbo will emerge

- Large sheet of foam core or cardboard (foam core sheets are 94-by-71 cm, or 37-by-28 in.)

- Ruler (marked in centimeters and/or inches)

- Scissors

- Pen or pencil

- Marker for drawing thick lines

- Cellophane tape

- Magazine or cardboard about 3 mm (1/8 in.) thick

- Household iron

- Transparent shipping tape

- 2, 10d gauge, 3 in. (7.62 cm) nails, weighing 7 gm each

- 2.5-by-2.5 cm (1-by-1 in.) chunk of Styrofoam to use as a cushion for nail heads

Assembly

1. You'll lay out the jumbo using measurements rather than using a template, because a template would be too large to handle. Figure 4-3 shows the parts of the jumbo's wing.

Make the first measurements from two opposite corners. Along the longer edge of the foam, make a mark 16 cm (6 5/16 in.) from the lower-right corner. Then, along the shorter edge, make another mark 11 cm (4 3/8 in.) from the corner. Use the marker to connect the marks with a thick line; this is where you'll make a cut. Do the same for the opposite (upper-left) corner. The foam sheet should look like Figure 4-4 when you're done.

FIGURE 4-3   Plan showing names of each part of a jumbo wing

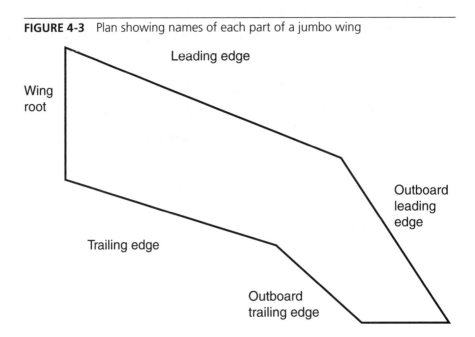

2. Now make another mark along the long edge of the sheet, 20 cm
(8 in.) from the lower-right corner. Draw a thin line connecting this
mark and the upper-right corner. Do the same for the upper-left corner
and lower-left corner. Your sheet should now look like Figure 4-5.

FIGURE 4-4   A thick line is drawn in opposite corners.

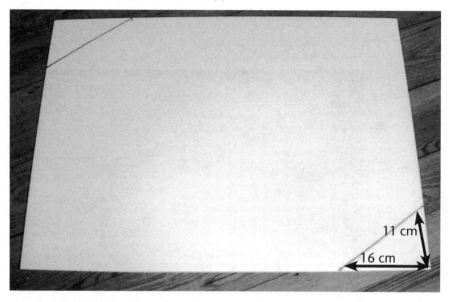

FIGURE 4-5   A thin line is drawn to the corners.

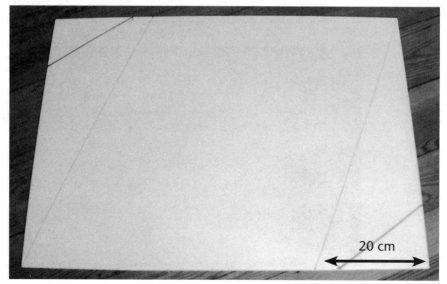

3. Draw a thick line atop the thin line, extending 26 cm (10 1/4 in.) from the upper-right corner. Do the same for the opposite corner, so the layout looks like Figure 4-6. You'll be cutting along this thick line.

4. Along the short edge of the foam sheet, make a mark 28 cm (11 in.) from the lower-right corner. This mark should appear at the location of the pen shown in Figure 4-7.

FIGURE 4-6 Thick line marking a cut line

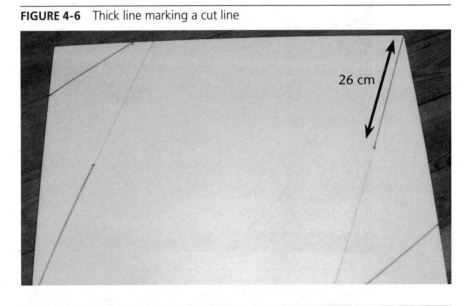

FIGURE 4-7 Mark the next location (shown at the pen).

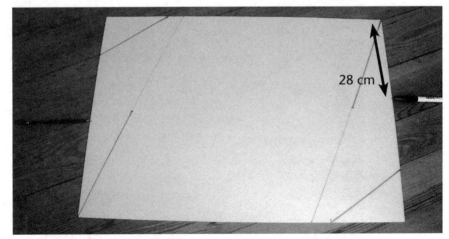

5. Draw a thin line connecting this mark and the end of the thick line you drew in step 3. Do the same for the opposite side, too (see Figure 4-8). (Notice that in the figure, the line extends past the end of the thick line; you'll see why in the next step.)

FIGURE 4-8   Thin lines on both sides

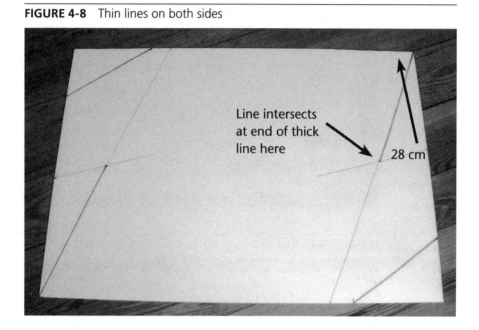

Line intersects
at end of thick
line here

28 cm

6. Now draw a thick line 22 cm (8 5/8 in.) from the edge of the Depron sheet, atop the thin line you made in step 5. Do this on both sides so the layout now looks like Figure 4-9.

7. The last lines you'll draw will connect the end of the outward trailing edge line and the wing root to mark the inward trailing edge. Start with thin lines here in case you mess up. Draw two parallel thin lines as shown in the middle of the sheet in Figure 4-10. These are the inward trailing edge lines for both wings.

FIGURE 4-9 Thick lines marking the outward trailing edge of each wing

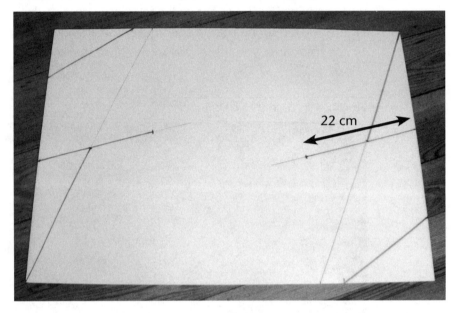

FIGURE 4-10 Thin lines marking the trailing edge of each wing

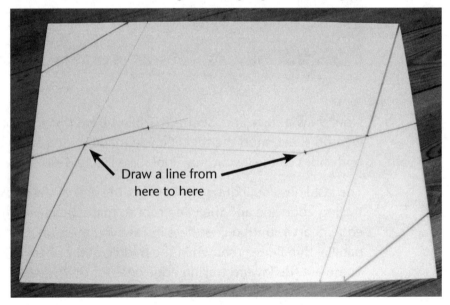

8. Finish by redrawing over these thin lines with thick lines, so the layout looks like Figure 4-11.

9. Now you'll start cutting the outer trailing edge of the wing. Begin cutting out the wings, at the point where the wing root line meets the lower-left corner. The two wings will be joined here, so it is important that this cut is smooth and straight (see Figure 4-12). Stop short of the next line intersection (where the thick line ends) to avoid cutting into the trailing edge of the other wing (Figure 4-13).

FIGURE 4-11   The finished layout marking cut lines for both wings of the jumbo

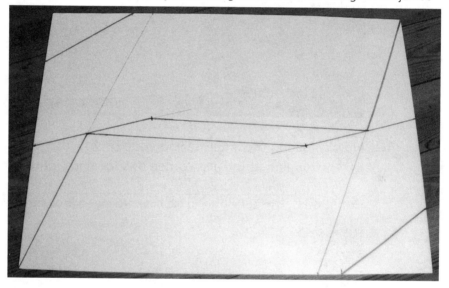

FIGURE 4-12   Cut out the wing from the jumbo layout.

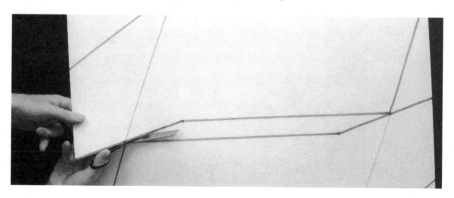

FIGURE 4-13   The wing root cut must be stopped slightly short of the next line intersection so as not to damage the other wing.

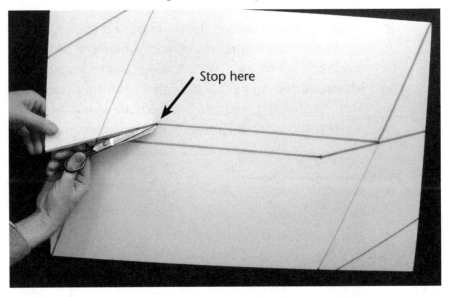

10. Now cut along the outer trailing edge of the other wing (see Figure 4-14), stopping at the first corner, as shown in Figure 4-15.

FIGURE 4-14   Start cutting along the outer trailing edge of the other wing.

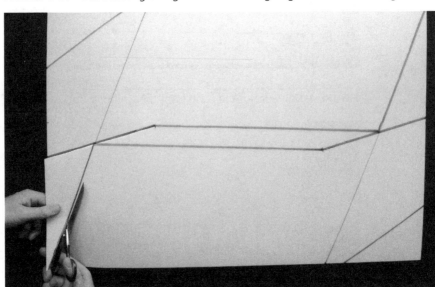

11. Cut along the line separating the leading and wing tip corner, as shown in Figure 4-16. At this point, the Depron sheet should look like Figure 4-17.

FIGURE 4-15   Stop cutting at the corner.

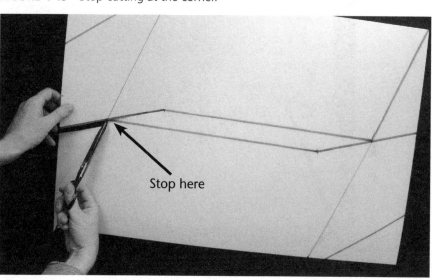

FIGURE 4-16   Cut off the corner and save this piece, because it will be used as part of the wing.

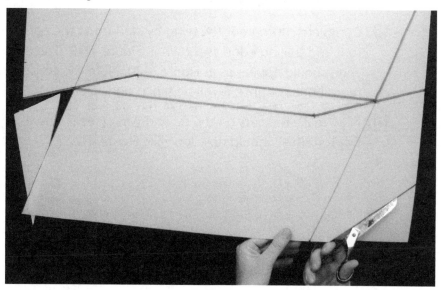

FIGURE 4-17   The jumbo wing emerges from the layout.

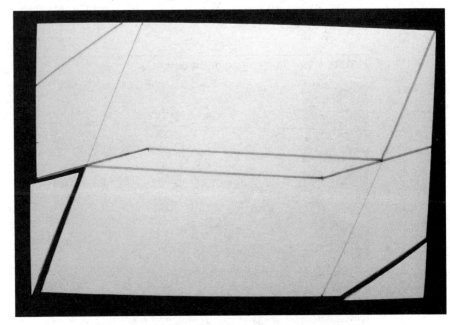

12. Start cutting at the opposite side of the sheet, as shown in Figure 4-18. Stop just short of the corner between the outboard trailing edge and the trailing edge.

13. Complete cutting out the wing by starting at the corner of the wing root and trailing edge as shown in Figure 4-19. The completed first wing should appear as in Figure 4-20. Note the corner has been repositioned (right). It will be taped here later.

14. Now make the cuts on the second wing (see Figure 4-21). The Depron sheet should now look like Figure 4-22.

FIGURE 4-18   Continue cutting along the trailing edge.

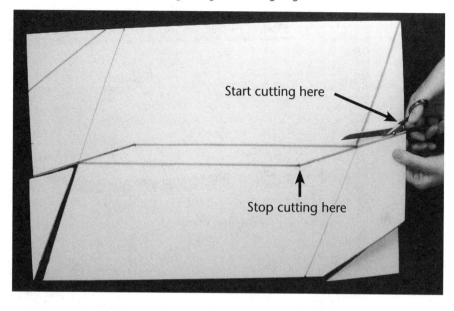

FIGURE 4-19   Last cut for the first wing

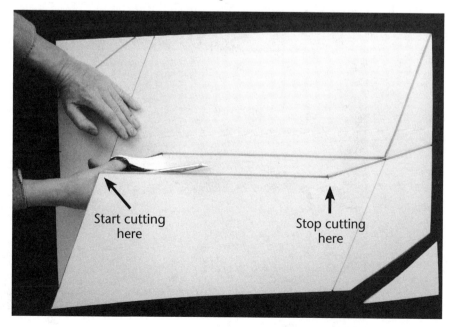

FIGURE 4-20 Completed cutout of the first wing; the corner piece has been repositioned to the location where it will be taped later (at lower right).

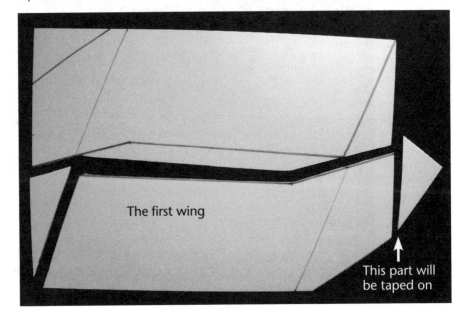

The first wing

This part will be taped on

FIGURE 4-21 Start cutting out the second wing.

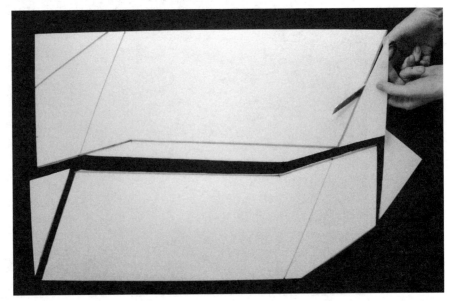

FIGURE 4-22 The Depron sheet after all cuts are made

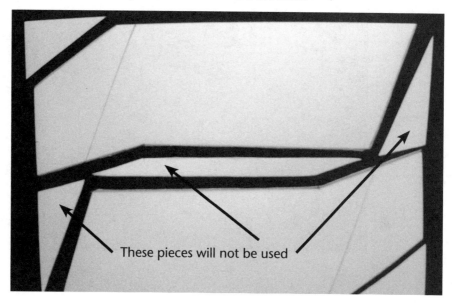

These pieces will not be used

15. Turn over one of the wings and its corresponding corner piece and join the two wings at the root to produce the final layout of the jumbo, as shown in Figure 4-23.

16. Make six marks on each wing root edge, measuring from the leading edge at these points, as shown in Figure 4-24. Make a mark 44 cm (17 3/8") from the wing root and drop a line to the corner on the trailing edge as shown in figure 4-25. Make the same marks on this line as you did on the wing root (see Figure 4-25). You'll draw lines between these corresponding marks to indicate where the thermoformed bends will be made to create the airfoil shape of the wings.

- 1.6 cm (5/8 in.)
- 3.25 cm (1 1/4 in.)
- 4.8 cm (1 7/8 in.)
- 6.5 cm (2 5/8 in.)
- 9.8 cm (3 7/8 in.)
- 13 cm (5 1/8 in.)

FIGURE 4-23 Layout of the jumbo cut parts

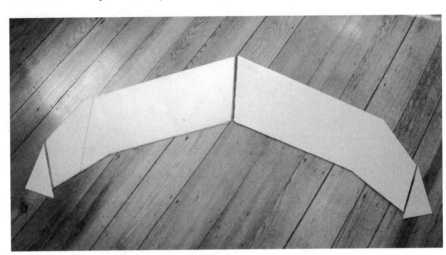

FIGURE 4-24 Marks made at the root edges

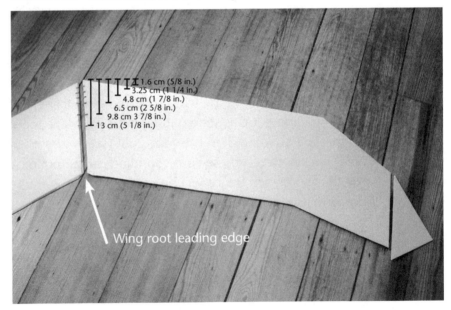

FIGURE 4-25   Mark the leading edge the same way you did on the wing root.

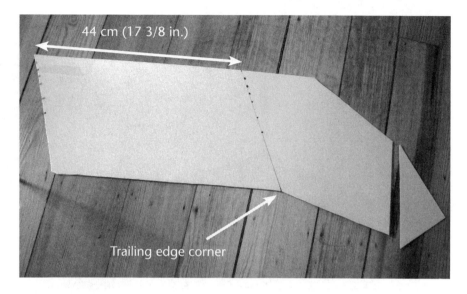

17. Connect each mark at the wing root and outer line, drawing six lines parallel to the leading edge. Do the same to the opposite wing so the airfoil lines appear as shown in Figure 4-26.

18. Turn over each wing and mark the positions for the elevon thermoform lines, 2 cm (3/4 in.) down from the leading edge wing tip corner and 2 cm (3/4 in.) up from the corner of the outboard and trailing edge as shown in Figure 4-27.

FIGURE 4-26 Lines parallel to the leading edge mark the thermoform bends that will serve as the airfoil.

FIGURE 4-27 Marks for the elevon thermoform line; notice that the elevon lines are on the opposite side of the wing from the airfoil lines.

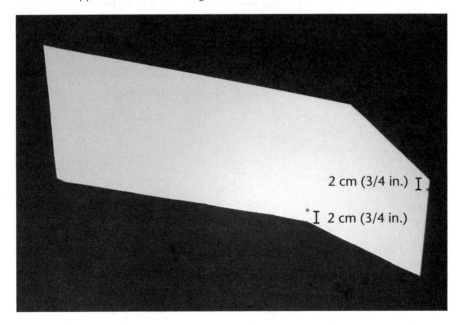

19. Join the marks to produce the first elevon thermoform line as shown in Figure 4-28.

20. Turn over each wing and tape the corner piece to the wing tip as shown in Figure 4-29. Both wings should now look as in Figure 4-30.

FIGURE 4-28 The forward-most elevon thermoform bend line on both wings

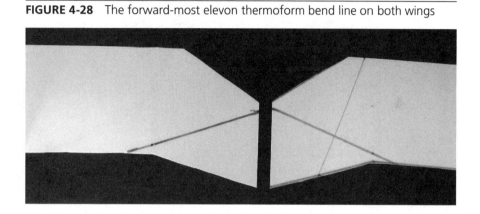

FIGURE 4-29 Tape the corner piece to the wing tip on the surface with the airfoil lines.

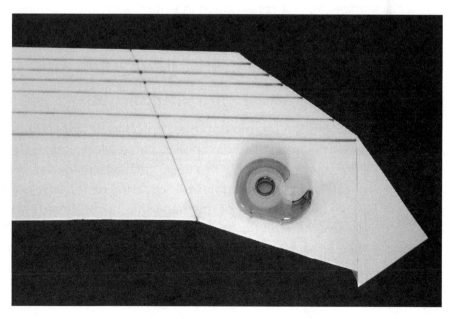

FIGURE 4-30 Both wings with the corner pieces taped on; the corner pieces will increase the effectiveness of the elevons.

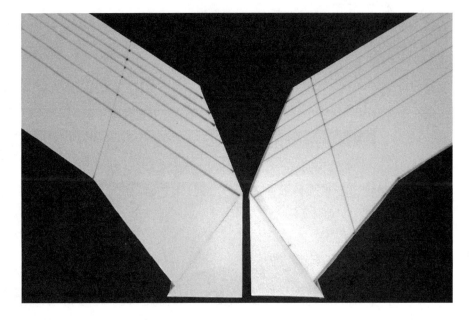

21. Turn over the wings again and complete the elevon bend lines by marking down from the first elevon line three more lines at the following points (as shown in Figure 4-31):

    - 3 cm (1 in.)

    - 6 cm (2 1/4 in.)

    - 10 cm (3 1/2 in.)

22. Now you can thermoform the Depron as described in Chapter 3. Set the iron for acrylic or silk. Align a wing along a straight edge about 3 mm (1/8 in.) thick, such as a magazine or a piece of cardboard, to thermoform each airfoil bend. Tape a sheet of paper over the cover of the magazine (or cardboard); you'll be drawing lines on this sheet to keep the magazine (or cardboard) in the correct position as the iron slides down each thermoform line. Draw a vertical line 1 cm (3/8 in.) from the edge of the magazine

FIGURE 4-31 Elevon thermoform lines. Notice how the lines extend onto the corner piece that was taped to each wing tip.

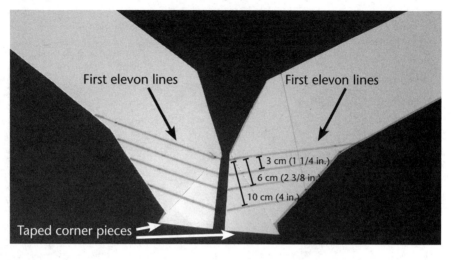

First elevon lines   First elevon lines

3 cm (1 1/4 in.)
6 cm (2 3/8 in.)
10 cm (4 in.)

Taped corner pieces

FIGURE 4-32 Draw a vertical line on the paper taped to the magazine (or cardboard) to place the magazine's edge about 0.5 cm (3/16 in.) from each thermoform bend line.

Magazine

(or cardboard), as shown in Figure 4-32. You'll use this line as a guideline for placing the edge of the wing as you thermoform each airfoil bend.

23. Place the wing over the magazine, with the wing's edge aligned with the line you drew on the magazine, as shown in Figure 4-33.

24. Use the straight part of the iron (the back part) to thermoform the Depron sheet into the groove created by the magazine's edge, as shown in Figure 4-34.

FIGURE 4-33   Positioning the wing over the magazine to create a thermoform bend along the line parallel to the leading edge

FIGURE 4-34   The hot iron is used to thermoform a bend along the line in the groove between the edge of the magazine and the table top.

25. Slide the hot iron to where the magazine ends (see Figure 4-35). The magazine is smaller than the wing, so you'll need to move the magazine to continue the bend.

26. Slide the magazine along the leading edge, overlapping with the thermoform bend and keeping the line parallel to the leading edge. Continue thermoforming the bend as shown in Figure 4-36 until a thermoformed bend has been created along the entire line to the end of the wing.

FIGURE 4-35 Keeping the magazine in place, slide the iron to where the magazine ends.

FIGURE 4-36   Slide the magazine down along the wing to create a continuous bend along the thermoform lines parallel to the leading edge of the wing.

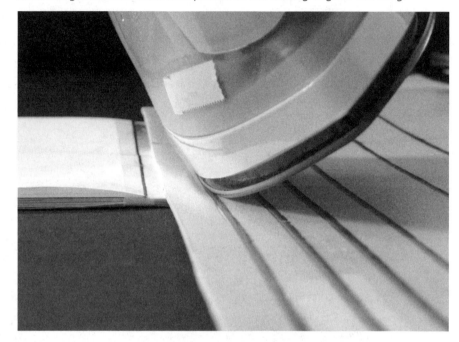

27. Draw another line on the magazine 2.8 cm (1 1/16 in.) from the edge of the magazine and parallel to the first line, as shown in Figure 4-37. Then line up the edge of the wing with this second line and get ready to thermoform the second airfoil bend.

28. You'll continue thermoforming a bend along each line in the same way for the rest of the lines, as shown in Figure 4-38. But first, you need to draw these lines. Starting with the third line (you drew the second one in step 27) add parallel lines:

- Line 3, 4.5 cm (1 3/4 in.) from edge of magazine

- Line 4, 6 cm (2 5/16 in.) from edge of magazine

- Line 5, 9 cm (3 1/2 in.) from edge of magazine

- Line 6, 11.5 cm (4 1/2 in.) from edge of magazine

FIGURE 4-37   Draw another line the length of the magazine 2.8 cm (1 1/16 in.) from the edge of the magazine (so as to position the edge of the magazine 0.5 cm up from the second thermoform bend line).

2.8 cm (1 1/16 in.)

FIGURE 4-38   Thermoform a bend along each line, using the position lines you marked on the paper taped to the magazine to help you position the wing.

11.5 cm (4 1/2 in.)
9 cm (3 1/2 in.)
6 cm (2 5/16 in.)
4.5 cm (1 3/4 in.)
2.8 cm (1 1/16 in.)
0.5 cm (3/16 in.)

After thermoforming all the bends in all the lines, the wing should look like Figure 4-39. It's starting to look like a real airplane wing!

FIGURE 4-39   The top of the wing after thermoforming bends in each line on the other side

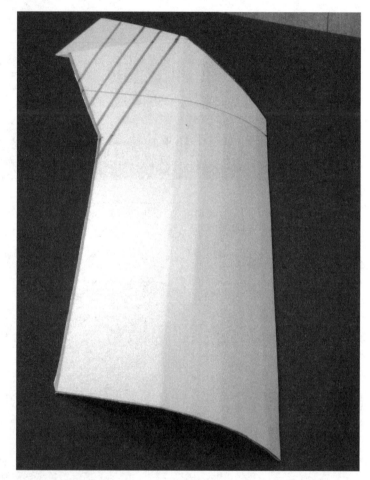

29. Now turn over the wing so the elevon thermoform lines show. Thermoform these lines in the same way, using the edge of the magazine, as shown in Figure 4-40.

30. Continue thermoforming each line. Figure 4-41 shows the magazine in position to thermoform the last elevon line. The wing should now look like Figure 4-42.

FIGURE 4-40   Thermoform the elevon bends. Because the elevon lines are shorter than the magazine, it is easier to position the elevon edges 0.5 cm (3/16 in.) from the magazine edge.

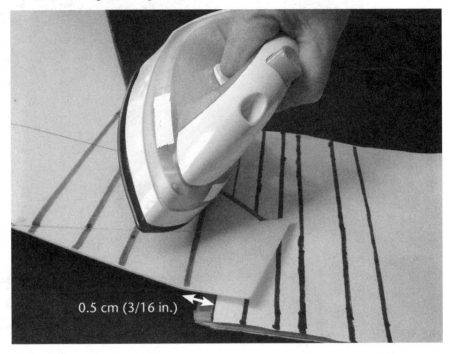

0.5 cm (3/16 in.)

FIGURE 4-41   The magazine in position to thermoform the last elevon line

FIGURE 4-42 The jumbo wing with thermoformed airfoil and elevon bends

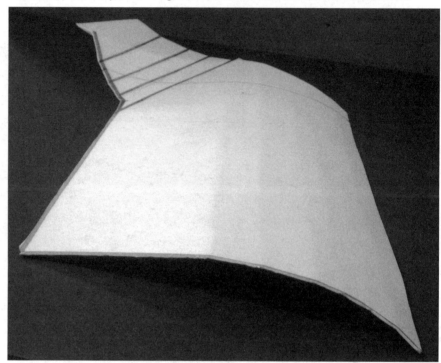

31. Next, you need to thermoform the wing washout angle into the wing in a similar way as how you did for the X-surfer in Chapter 3. Holding the hot iron a couple of centimeters above the table, twist the leading edge of the wing up so the top surface of the wing comes into contact with the iron, as shown in Figure 4-43.

32. Slide the hot iron toward the wing tip while keeping the leading edge twisted up (see Figure 4-44).

33. Turn over the wing and thermoform an extra airfoil bend from midway between the last airfoil bend and trailing edge and the middle of the trailing edge, as shown in Figure 4-45. The same process can be used to thermoform the airfoil, elevon bends, and wing washout for the other wing. The finished wing twist should look as in Figure 4-46.

FIGURE 4-43   Thermoforming wing twist (washout) into the wing

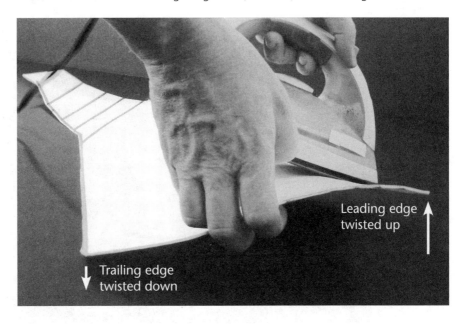

Leading edge
twisted up

Trailing edge
twisted down

FIGURE 4-44   Move the hot iron down the wing while you twist the leading edge upward.

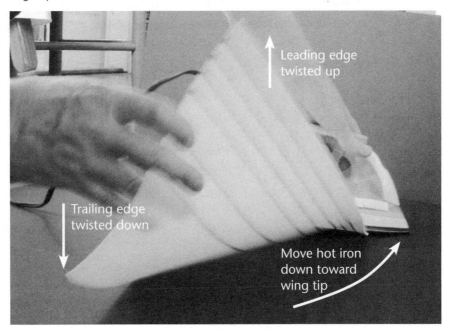

Leading edge
twisted up

Trailing edge
twisted down

Move hot iron
down toward
wing tip

FIGURE 4-45 Thermoform an extra airfoil bend into the trailing edge of the wing.

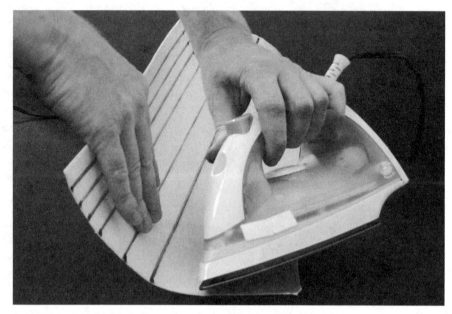

FIGURE 4-46 Thermoformed wing washout

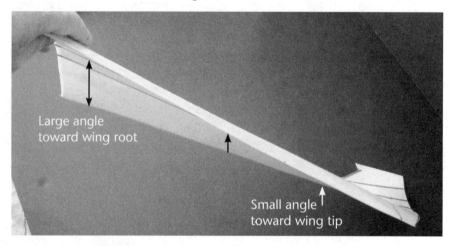

Large angle toward wing root

Small angle toward wing tip

34. After you've thermoformed both wings, airfoils, elevon bends, and washout, you can attach the wings at the wing root at three attachment points: one semi-permanent attachment at the high point of the airfoil on the top side, and two permanent attachments at the nose and trailing edge on the underside. With this arrangement, the two bottom attachments act as hinges when the top attachment is removed, so the wings can fold to half the normal wingspan for storage, as shown in Figure 4-47.

35. Attach a 9 cm (3 1/2 in.) length of shipping tape to anchor permanently to the Depron surface. Start by attaching 1 cm (3/8 in.) to the underside, wrapping the tape across the wing root edge and the rest extending along the top surface. Do this to the other wing, trying to align the tape evenly on both wings so the top side looks like Figure 4-48.

FIGURE 4-47   Completed jumbo with wings folded for storage

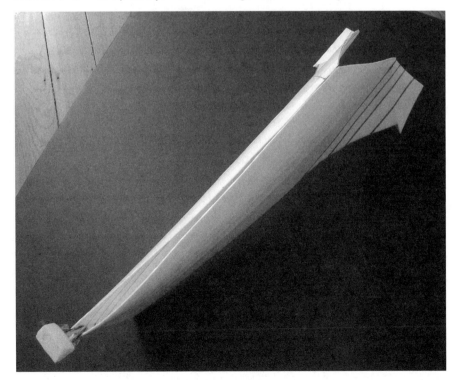

FIGURE 4-48 The anchor tape on both wings at the high point of the airfoil on the top of the jumbo

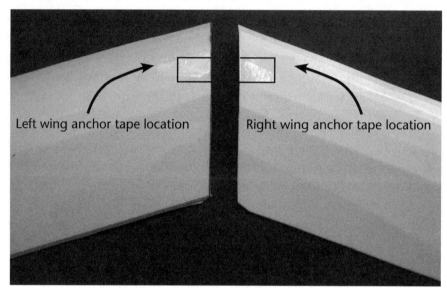

Left wing anchor tape location   Right wing anchor tape location

36. Turn over the wings and tape the undersides together, holding the two wings as close together as you can. Leave a little bit of tape over the edge, as shown in Figure 4-49. This tape will be wrapped around the leading edge and smoothed over the top of the wings.

37. Now use cellophane tape to attach the trailing edge wing roots, keeping the wings as close together as you can. The tape should look like Figure 4-50.

<u>NOTE</u>   *Cellophane tape is used here instead of shipping tape for two reasons: first, cellophane tape is lighter, and second, this joint should come apart first in a crash to help relieve stress on the jumbo's other attachments.*

FIGURE 4-49   Attach the wings at the nose, leaving a bit of tape to wrap over the top surface of the wings.

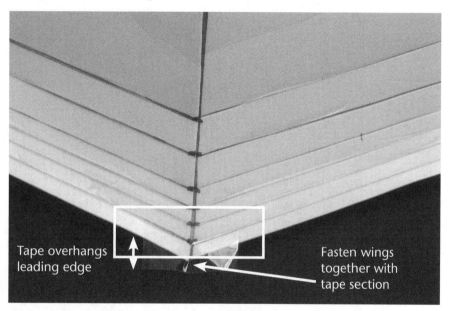

FIGURE 4-50   Attach the trailing edge of the wings using cellophane tape.

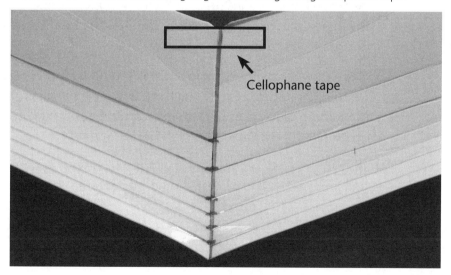

FIGURE 4-51   Make indentations in the Styrofoam block.

38. Make indentations in the little Styrofoam block using the two nail heads, as shown in Figure 4-51. The nail head will serve as nose weight.

39. Tape the nails to the Styrofoam block using a long enough length of shipping tape to extend halfway down each nail, around the Styrofoam block, and back to itself on the underside, even with the beginning of the tape (Figure 4-52).

40. Attach the nose weight to the wings with shipping tape, as shown in Figure 4-53. The weight should be centered as closely as possible for now.

FIGURE 4-52 Insert the two nail heads into the Styrofoam block and attach them with shipping tape.

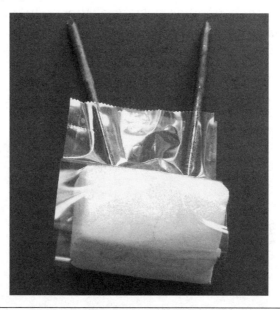

FIGURE 4-53 Attach the nose weight to the wings.

41. Finally, attach the top of the wings together with a length of shipping tape, as shown in Figure 4-54. Make the length of the tape shorter than the tape anchoring to the Depron surface so when the tape is removed it will not pull the anchor tape with it. Also folding over a small edge of the tape will make removing the tape easier.

FIGURE 4-54   Attach the wings at the high point of the airfoil.

Trim the Jumbo

The jumbo is now ready to be trimmed. First trim the nose weight by making a test glide and noting whether the jumbo glides smoothly, dives, or slows and then dives. Launch the jumbo at a fast walking speed with the nose pointed a little downward. If the glider dives, try moving the nose weight in (toward the wings), as shown in Figure 4-55. If the glider does not fly smoothly but pitches up and down or slows and then dives, the nose weight should be moved out a small distance from the wings.

FIGURE 4-55   This glider dove in a test flight so the nose was reattached inward to correct the problem (compare to Figure 4-53).

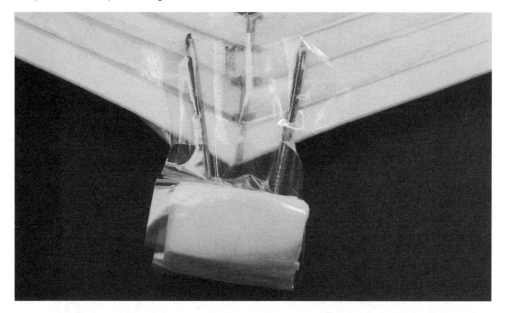

Once the jumbo is flying slowly and smoothly, sight from the front of the jumbo to reveal any asymmetries between the wings. For example, the front view in Figure 4-56 shows that the left elevon is deflected more than the right elevon. See Figure 4-57 for the corresponding glide when the left elevon has more upward deflection and the resulting left turning tendency.

FIGURE 4-56   Front view of the jumbo, showing asymmetries between the right and left elevons (here the left elevon is deflected a bit more upward than the right elevon)

FIGURE 4-57 Test glide of the jumbo with the left elevon up higher than the right elevon; the jumbo turns to the left

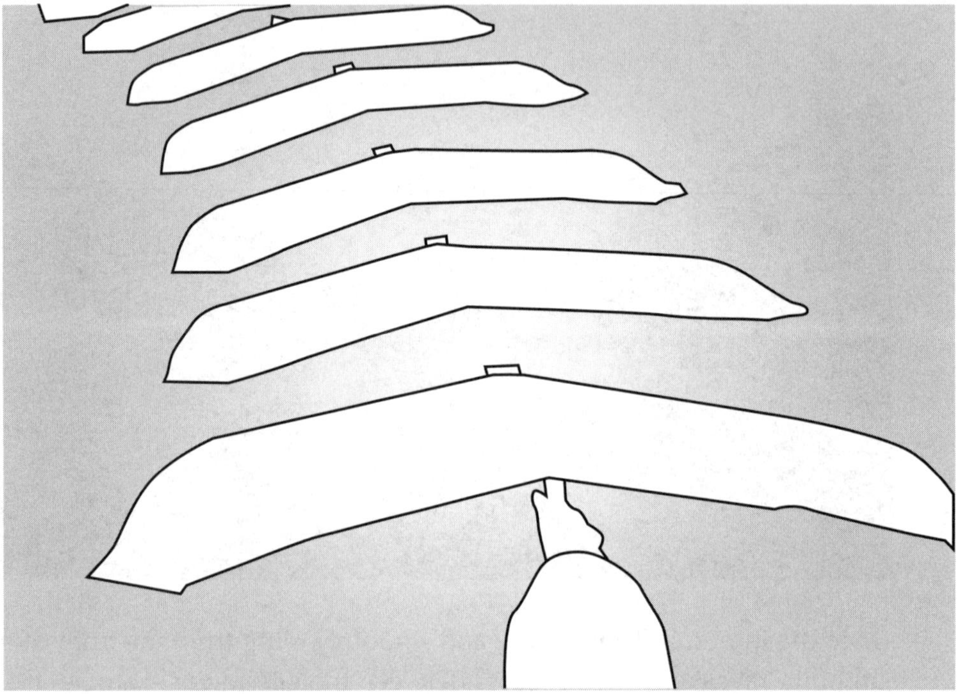

Sustain the Jumbo in Flight

The jumbo is flown in much the same way as the X-surfer from Chapter 3, but it will react much slower to the position of the paddle because of its size. Without the paddle, the glider glides down pretty quickly, as shown in Figure 4-58.

To sustain the glider in flight, launch it from as high as your arm can reach at a fast walking speed and with the nose pointed a little downward. Holding the paddle, start moving with the glider and gently position the top edge of the paddle under the trailing edge of the jumbo (see Figure 4-59).

FIGURE 4-58   The jumbo gliding without help from the paddle

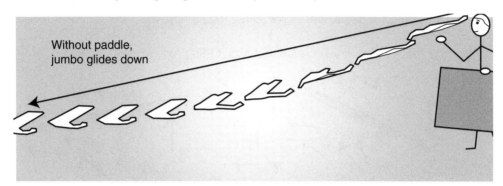

Without paddle,
jumbo glides down

FIGURE 4-59   Sustaining the glider with the paddle

With paddle generating updraft,
jumbo stays up

Also, moving the paddle forward or backward relative to the glider will help keep its nose at the correct angle. For example, if the nose points too far downward and the jumbo is picking up speed, moving the paddle forward a bit will raise the nose and pull the glider out of the dive (the equivalent of the airline cockpit alarm sounding "Pull up!! Pull up!!"). See Figure 4-60 for a diagram of how the forward or backward positioning of the paddle affects the angle of the nose of the jumbo.

FIGURE 4-60 The angle of the nose can be controlled by moving the paddle forward or backward as the jumbo glides. To pull out of a dive, move the paddle forward. If the jumbo slows too much and the nose angles up too much, move the paddle backward relative to the glider.

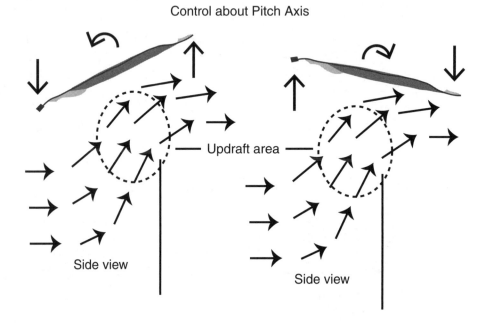

Control about Pitch Axis

Updraft area

Side view

Side view

Steer the Jumbo

Turning the jumbo will take more space than the X-surfer. To help with turning the jumbo, move the paddle toward the opposite side of the turn, as shown in Figure 4-61.

Figure 4-62 shows the jumbo in an S-turn maneuver. If you look closely at the beginning of the first turn, the glider is actually flying sideways. I'll talk about why that happens in the "Theory" section next.

Figure 4-63 demonstrates a 180-degree turning radius of about 6 meters (18 ft.). The dimensions of the area for this maneuver are 11-by-14 meters (35-by-46 ft.).

FIGURE 4-61   Turning the jumbo by moving the paddle to the outside of the turn

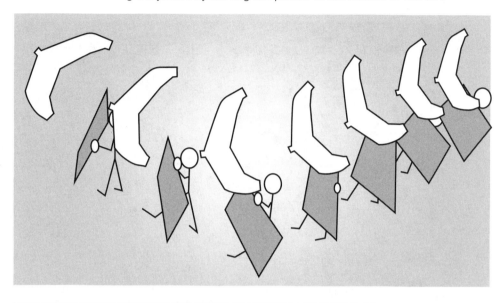

FIGURE 4-62   Jumbo executing an S-turn maneuver

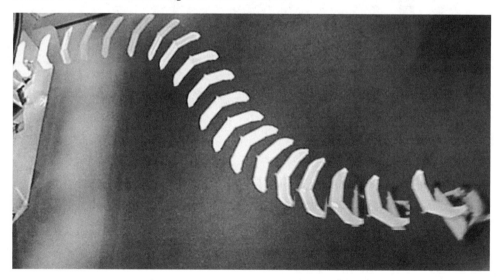

FIGURE 4-63   Jumbo executing a 180-degree turn in an area of 11-by-14 meters (35-by-46 ft.)

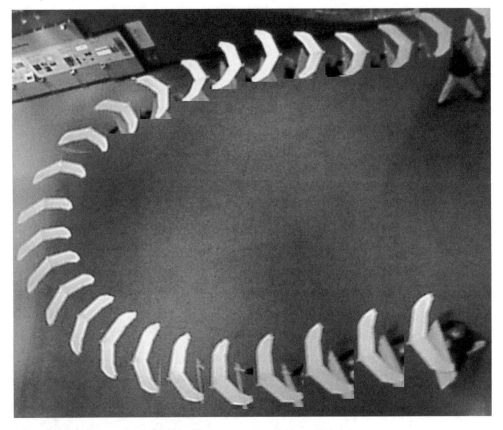

Theory

Why does the jumbo fly sideways at the beginning of a turn? When you start a right turn, the left wing must move higher. To make the left wing rise higher, you put more lift and thus higher drag on the left wing, which makes the jumbo point briefly to the left until the wing stops moving up. Although the glider is pointing to the left, it continues to fly straight or to the right and thus flies sideways (see Figure 4-64).

FIGURE 4-64   Jumbo executing a right turn; because the left wing must be lifted, it experiences increased drag, which makes the glider fly sideways to the right at the beginning of a turn.

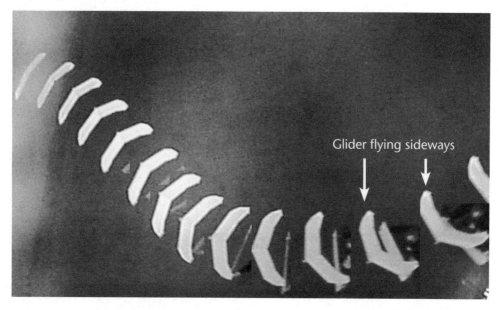

Glider flying sideways

Full-size airplanes, especially those with longer wings, experience this effect, called "adverse yaw," in a turn. In an airplane with onboard controls, adverse yaw is counteracted by using the rudder to coordinate the turn. Because the walkalong glider does not have a controllable rudder, it experiences a delay before each turn. This delay is most pronounced with the jumbo because it has a long wingspan. Flying the jumbo requires anticipation of the slow reaction time of the glider to your paddle movements. This is the difference between light and maneuverable gliders such as the tumblewing and a larger, heavier glider such as the jumbo.

5

Butterfly Glider

Project size: Small   **Skill level:** ★★★★

In this chapter, we'll mount a butterfly in such a way that it can be used as a walkalong glider (see Figure 5-1).

For this project, I encourage you to find a butterfly that is already dead, rather than catch or buy one. But if you're unable to find a dead butterfly, you can order one online from a respectable source such as www. butterfliesandthings.com. Butterflies can also be obtained from butterfly mounting kits.

Butterflies have life spans of not more than two weeks. Because of this, any butterfly garden needs a steady supply of larva to replace expiring butterflies. Butterfly gardens in nontropical areas that have tropical species are required to freeze any material being removed from the garden to prevent the spread of tropical parasites and diseases. So you could also ask a local garden if they have frozen butterflies available.

<u>NOTE</u>   *Yes, this project uses a real, dead butterfly. Although butterfly collecting and making displays are a common hobby, I realize that some readers might object to using a dead butterfly as a glider and skip this project. I respect that the choice is purely up to you.*

FIGURE 5-1   Mounted butterfly trimmed to fly as a walkalong glider

This project requires that you spread out the butterfly's wings and allow the insect to sit for a couple weeks before you can use it for a glider. Keep this in mind when you start this project.

What you'll need:

- Butterfly collected from source such as www.butterfliesandthings .com (the butterflies may come with wings folded so just follow the directions on how to spread the wings at the site; morpho, owl, and monarch butterflies are the best species for using as gliders)

- Flat piece of Styrofoam, 30-by-30 cm (12-by-12 in.)

- 12 thumbtacks

- 6 strips of paper, 10-by-2 cm (4-by-3/4 in.)

- Clear fingernail polish

- Dowel or pencil, 6 mm (1/4 in.) diameter

<u>NOTE</u>   *The butterfly suppliers at www.butterfliesandthings.com have special licenses and must follow specific regulations to purchase and sell butterflies.*

Assembly

1. Place the butterfly bottom side up on the Styrofoam, as shown in Figure 5-2. If your specimen has its wings folded, you will need to gently spread the wings. There are instructions for doing this at www.butterfliesandthings.com. When positioning the forewings, align the leading edges of the wing about even with the head. For display, the forewings are often positioned too far forward and would result in a center of lift too forward relative to the center of weight (see the Theory section in Chapter 3).

2. The forward wings are held flat against the board using the thumbtacks and paper strips. Slide a paper strip between the front and back wings, as shown in Figure 5-3. Use thumbtacks to tack down the paper at either end. Take care that no part of the thumbtack crushes the wings.

FIGURE 5-2 Place butterfly on the Styrofoam. Note the forward wings are swept back, about even or slightly behind the head.

Wings even with or slightly behind head

FIGURE 5-3 Slide a paper strip between the front and back wings.

3. Do the same for the other wing, as shown in Figure 5-4.

4. Gently adjust the position of the wings relative to the insect's body, making the leading edge of the wings about even with the head. Continue applying strips to hold the front wings flat out to the wing tips (see Figure 5-5).

FIGURE 5-4   Both front wings are held flat against the board by paper strips and thumbtacks. Leave sufficient space between the head and the edges of the paper strips to minimize the chance of fingernail polish getting between the butterfly's wings and the paper strips.

Allow space between head and edge of paper strip

5. Slide a dowel under the back wings to create a bend in the back wings, as shown in Figure 5-5. This bend creates an increased angle to the forward half and an elevator effect to the back half of the back wings.

6. Apply paper strips fastened down by thumbtacks to hold the back wings in the bent position over the dowel (see Figure 5-6).

FIGURE 5-5   Slide a dowel under the back wings.

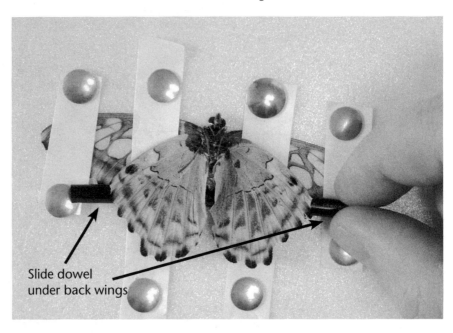

Slide dowel
under back wings

FIGURE 5-6   Paper strips hold the back wings in a bent position over the dowel.

Hold down back wings
with paper strips

FIGURE 5-7   Apply fingernail polish to strengthen the joints where the wings meet the body.

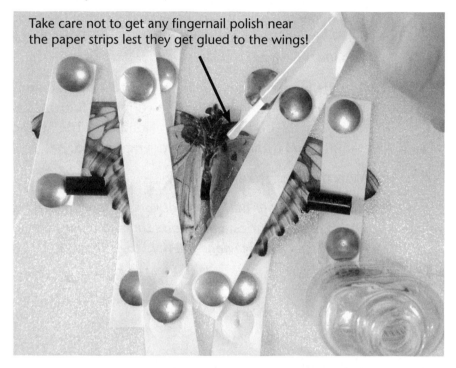

Take care not to get any fingernail polish near the paper strips lest they get glued to the wings!

7.   Apply some clear fingernail polish where the wings meet the insect's body to strengthen this area (see Figure 5-7).

   <u>NOTE</u>   *Bear in mind that any time you add fingernail polish it will increase the weight of the butterfly. Try and keep this added weight as far forward as possible; this will keep the butterfly more stable in flight.*

8.   Add fingernail polish to the insect's legs to strengthen them and create a small handle for handling the butterfly.

   <u>NOTE</u>   *Take care not to get fingernail polish between the butterfly and the board, because this will glue the insect to the board.*

9.   Allow the butterfly to dry for two weeks.

10. After a couple weeks, carefully remove the paper strips holding the back wings over the dowel and slide the dowel sideways out from under the back wings. Remove the rest of the paper strips, sliding them sideways from under the back wings. Be careful not to damage the wings.

11. Turn the butterfly right side up and further strengthen the wing attachment to the insect's body by applying fingernail polish to the top side of the body, as shown in Figure 5-8. The head can also be strengthened by adding some fingernail polish.

12. Allow the polish to dry for a day.

    <u>NOTE</u>   *Store your butterfly glider on the foam board with the same system of paper strips, dowel, and thumbtacks used to form the wings. This way, the wings will hold their shape.*

FIGURE 5-8   Strengthening the wings' attachment to the body by applying fingernail polish to the top of the body.

Trimming Your Butterfly Glider

Perform a test flight. If the butterfly glider slows and then dives, you can add more weight to the insect's head by adding some fingernail polish. If the butterfly glider turns to one side, the forward wings may need to be spread out against the Styrofoam board for additional time. Once the butterfly glider is gliding smoothly and slowly without turning too much, you are ready to try sustaining it with the paddle you made in Chapter 1.

Flying Your Butterfly Glider

You'll fly the butterfly glider in much the same way that you flew the paper airplane surfer in Chapter 2. Figure 5-9 shows a butterfly glider gliding without the paddle.

FIGURE 5-9   Butterfly glider flight trajectory without using the paddle to sustain it

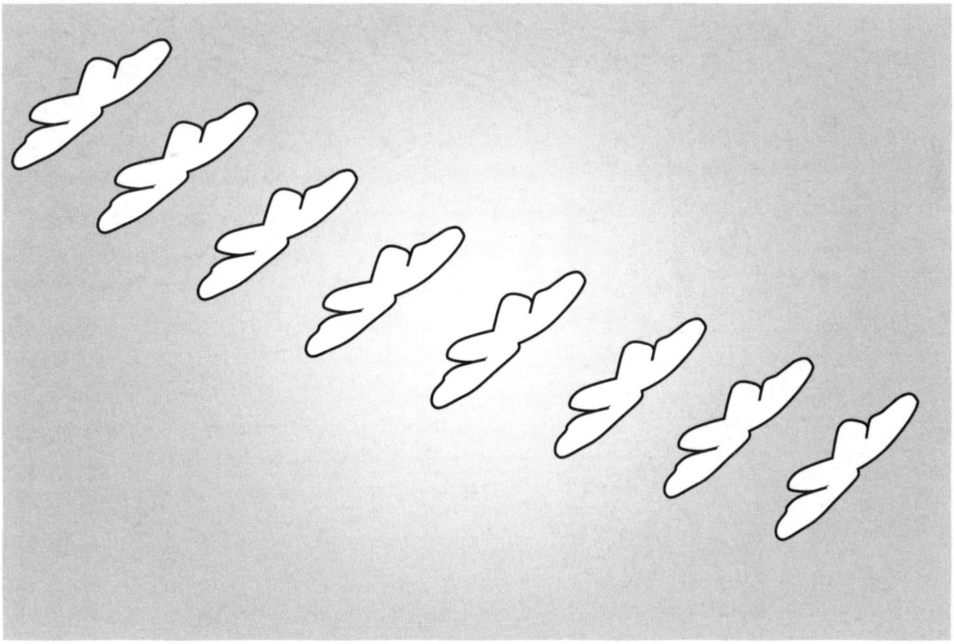

Now let the butterfly glider go from as high as you can and follow it with the paddle, placing the top edge of the paddle as close to the butterfly glider as you can. Move a little faster without having the butterfly glider travel over the paddle. Figure 5-10 shows the butterfly glider trajectory being sustained by the paddle.

By putting the paddle even closer to the butterfly glider, you can get it to climb higher, as shown in Figure 5-11.

Figure 5-12 shows the path of a butterfly glider in a turn, and then straight and level.

FIGURE 5-10   Flight trajectory of the butterfly walkalong glider being sustained with the paddle

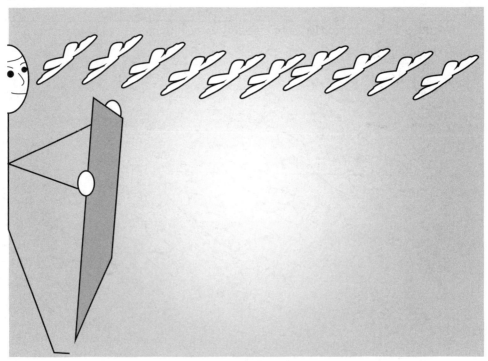

FIGURE 5-11   Flight trajectory using the paddle to make the butterfly glider climb

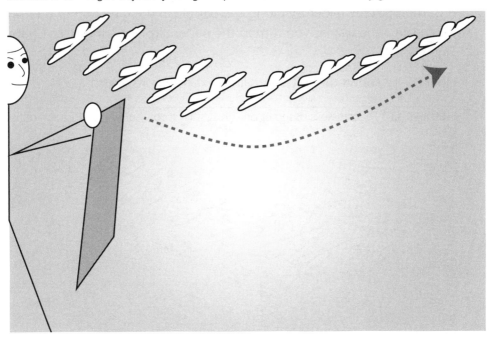

FIGURE 5-12   Butterfly glider overhead view of flight trajectory of turn then straight and level

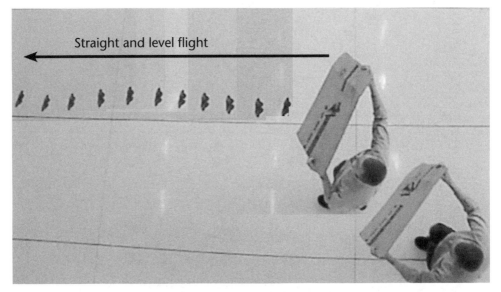

Straight and level flight

Figure 5-13 shows the butterfly glider turning to avoid the wall. You can turn the butterfly glider by tilting the paddle in the direction of the turn , in much the same way as you turned the paper airplane surfer (see Chapter 2, Figure 2-54).

Figure 5-14 shows the butterfly glider executing an S-turn.

FIGURE 5-13   The paddle is used not only to sustain flight, but also to turn the butterfly glider.

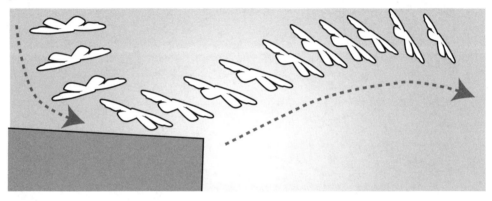

FIGURE 5-14   Overhead view of butterfly glider flight trajectory executing an S-turn

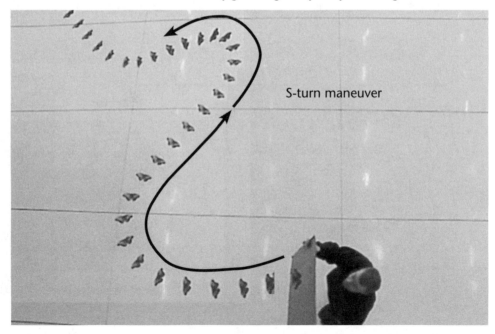

S-turn maneuver

Theory

Many plants and animals are well adapted to use air currents and the air as a means of locomotion. The butterfly's wings are especially adapted to the aerial environment. Butterflies mounted for show have as much of the wings showing as possible, and the front wings are far forward of the body. If you have the chance to observe live butterflies, you will see they hold the leading edge of their wings about even with their head as they rest on a leaf with wings spread flat. If a sudden gust of wind hits the insect, they want to be ready in a flying position. We mount the butterfly in this more natural position.

What other ways could you use natural materials to create a glider?

<u>NOTE</u>   *Wouldn't it be fun to fly in an updraft strong enough to hold you up? This is exactly what you can do in the indoor skydiving experience, SkyVenture. Learn more about this fascinating way of gliding with your own body at www.skyventure.com. Because your body would not be very efficient at generating lift, you'll need to learn how to stabilize yourself in flight and how to maneuver using just the wind.*

6

Baby Bug

Project size: Medium   **Skill level:** ★★★★

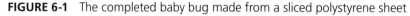

What would happen if you used an even lighter foam material to make a walkalong glider? The baby bug (see Figure 6-1) is made from such a material: thin-sliced, expanded polystyrene. Thin polystyrene is a very light material that makes it easy to form the airfoils and elevons of the baby bug design.

The baby bug flies almost as slow as a tumblewing, but it has a much better glide ratio and can even be flown without a paddle. And if you use a paddle, the slow flying speed reduces the turbulence produced by the paddle, which keeps the airspace more flyable. The baby bug is so much fun to fly!

FIGURE 6-1   The completed baby bug made from a sliced polystyrene sheet

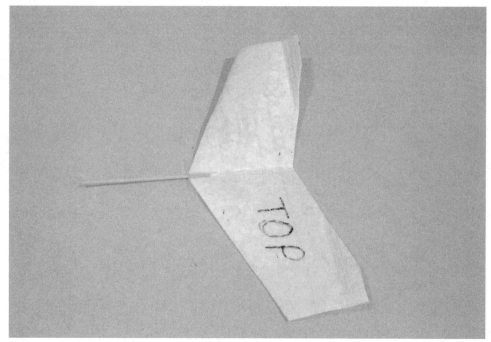

NOTE   *As of the writing of this book there are several possibilities where ready-made, thin-sliced polystyrene foam might be available for purchase. Obtaining the sheets pre-cut eliminates the need to construct a hot wire cutter and the danger of harmful vapors and getting burned from the heat of the hot wire. Please check the web site www.walkalongglider.info for up-to-date information on sources of suitable thin-sliced (0.5 mm thick) foam.*

The baby bug design and the design for a rudimentary hot wire cutter come from Slater Harrison, a school teacher from Williamsport, Pennsylvania. The hot wire cutter can be used to make thin sliced foam sheets instead of purchasing them. We also have to thank Michael Thompson, a college student from Wisconsin, for advancing the concept of using thin-sliced polystyrene for walkalong gliders.

What you'll need (Figure 6-2):

- Paper for the template
- Pen or pencil

FIGURE 6-2   Supplies to make the baby bug

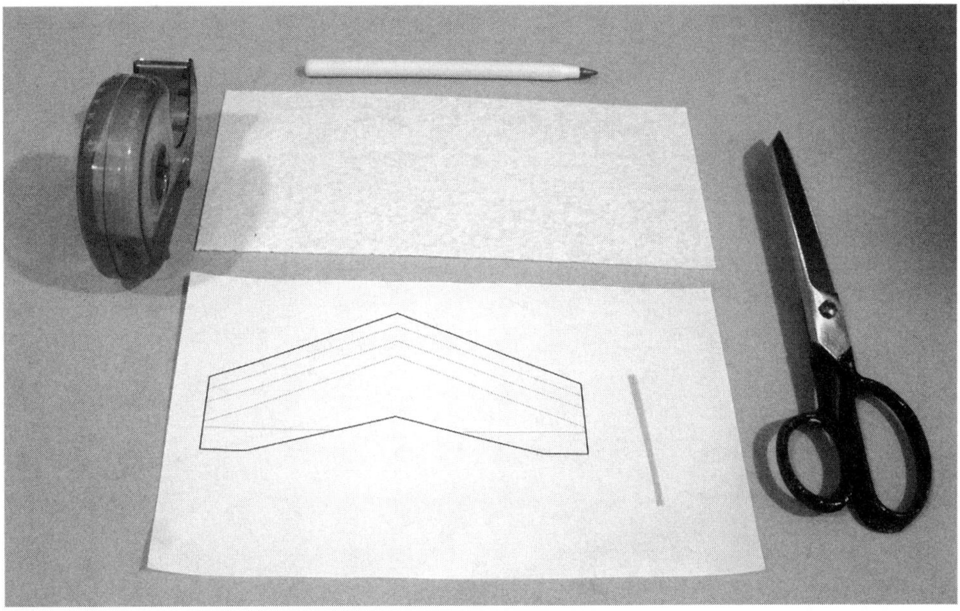

- 0.6 mm thin-sliced expanded polystyrene sheet, 7.5-by-12.5 cm (3-by-5 in.)

- Scissors

- Cellophane tape

- Book about 2.5 cm (1 in.) thick

- Flat toothpick

Assembly

1. Print the baby bug design available for download from www. walkalongglider.info and shown in Figure 6-3. The wingspan pattern should be scaled to 10 cm (4 in.).

2. "Rough cut" the design template, leaving about 0.5 cm (1/8 in.) of paper outside the thick line, as shown in Figure 6-4.

3. Tape the template to the sliced foam sheet, as shown in Figure 6-5, using two pieces of tape on the wing tips.

FIGURE 6-3 The baby bug template

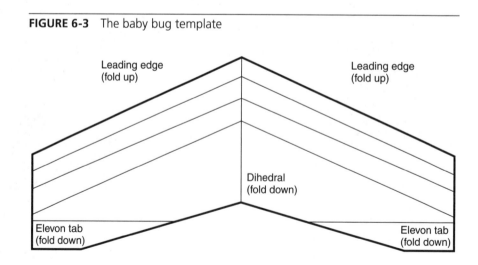

FIGURE 6-4   Rough cut the baby bug template.

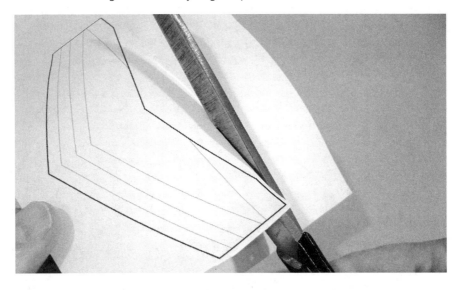

FIGURE 6-5   Tape the template to the sliced foam sheet.

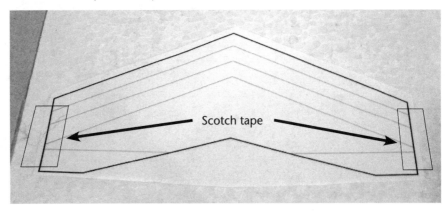

4. Cut off some of the excess foam, leaving about 2 cm (1/2 in.) of foam beyond the tape on the wing tip, as shown in Figure 6-6.

5. Cut along the trailing edge of the design, as shown in Figure 6-7. When you're done, the foam sheet should look like Figure 6-8.

FIGURE 6-6 Cut the excess foam from the wing tip area.

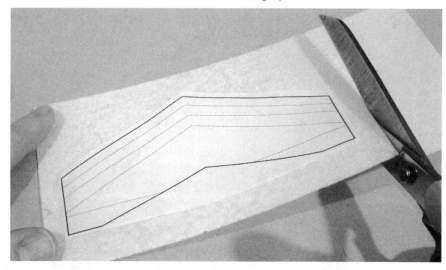

FIGURE 6-7 Cut the trailing edge of the foam sheet, leaving the wing tip uncut so the tape will hold the template to the foam.

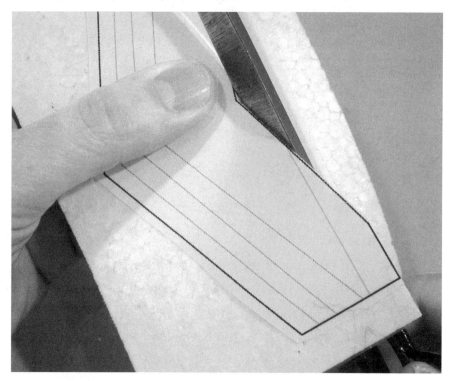

FIGURE 6-8   Cut the trailing edge on the template's thick line, leaving the wing tips uncut.

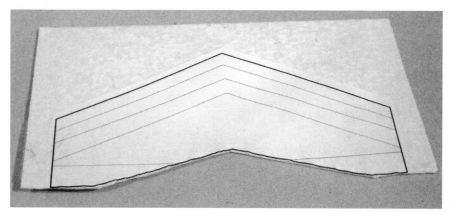

6.   Do the same for the leading edge, leaving the wing tip uncut, as shown in Figure 6-9. When you're done, the foam should look like Figure 6-10.

FIGURE 6-9   Cut the leading edge on the thick line, leaving the wing tip areas uncut.

FIGURE 6-10 The foam slice sheet with leading and trailing edges cut

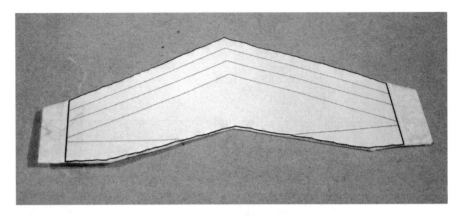

7. Align the thin elevon lines with the edge of a book, as shown in Figure 6-11.

FIGURE 6-11 Align the thin elevon lines with the edge of a book.

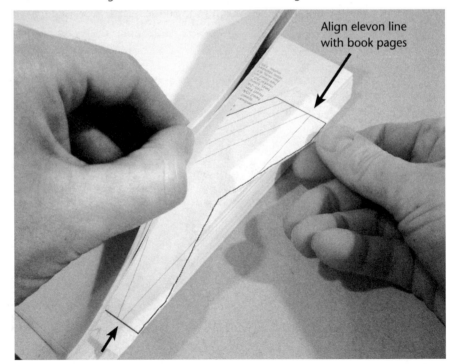

Align elevon line
with book pages

8. Close the book so that the elevon tabs are showing (Figure 6-12).

9. Press down on the elevon tabs to start the fold, as shown in Figure 6-13. Figure 6-14 shows the tabs after they've been folded.

FIGURE 6-12   Close the book so you can create a straight fold along the elevon line.

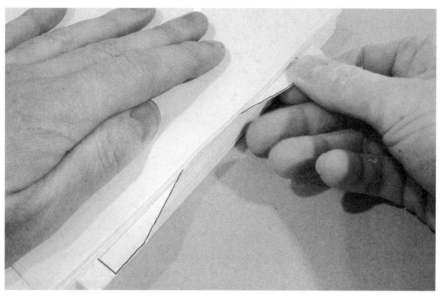

FIGURE 6-13   Fold down both elevon tabs.

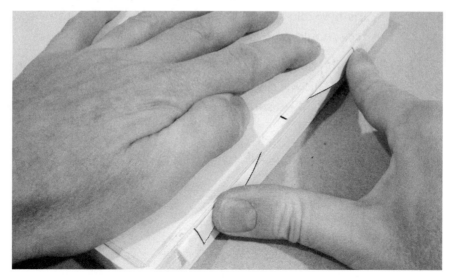

FIGURE 6-14   The elevon tabs after using the book to create a straight edged fold

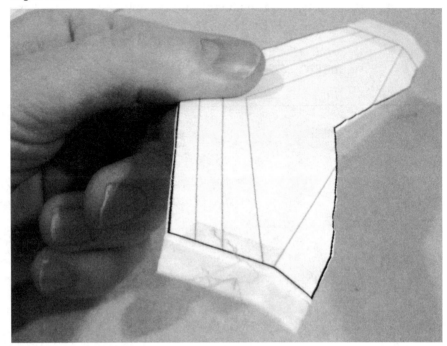

10. Pinch the elevon tab to increase the folds, as shown in Figure 6-15; the new fold should be about 45 degrees. The top view of the foam slice should now look as shown in Figure 6-16.

11. Use the book again to fold the centerline fold, as shown in Figure 6-17. This will give each wing an upward angle, called *wing dihedral*. Make sure the design paper is face up before you make the fold.

FIGURE 6-15   Fold the elevon tabs a bit more.

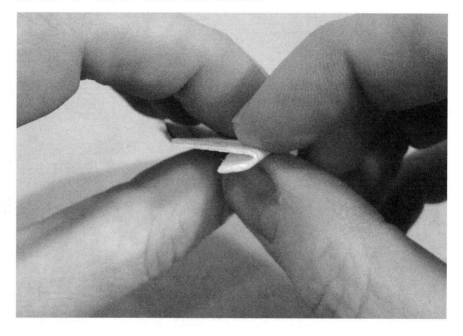

FIGURE 6-16   Top view of the foam slice with elevon tabs folded

FIGURE 6-17 Use the book to fold the centerline fold to give the wings an upward angle.

Align centerline fold with edge of book pages

12. Fold the foam down, as shown in Figure 6-18.

13. Increase the centerline fold, as shown in Figure 6-19. The foam should now look like Figure 6-20. Label the top so you'll know which side is up when the template is removed.

FIGURE 6-18   Fold the foam down against the edge of the book.

FIGURE 6-19   Increase the centerline fold.

FIGURE 6-20   Top view of the foam slice after folding the elevon and centerline

FIGURE 6-21   Reduce the folds by gently pressing on the bottom side of each fold.

14. Reduce the severity of the folds by gently pressing each fold from the bottom side, as shown in Figure 6-21.

15. Roll the round barrel of a pen over the right wing to create a rounded airfoil shape. Align the third thin line in from the leading edge with the book edge, and roll the pen gently to form a small rounded fold between the second and third thin lines, as shown in Figure 6-22. Be gentle when making this fold, because it should be the least angled of the next couple of folds.

16. Align the second thin line in from the leading edge with the book edge and roll the pen a little more firmly and a few more times between the first and second lines to continue shaping the airfoil, as shown in Figure 6-23.

17. Do the same between the leading edge and the first thin line to form a larger angle to the round fold than the last fold, as shown in Figure 6-24. The right wing should now have a rounded fold with increasing angle from the third thin line to the leading edge, as shown in Figure 6-25 (the right wing appears at left because the gilder is upside down).

FIGURE 6-22 Roll the barrel of the pen on the foam to form a rounded fold between the second and third thin lines in from the leading edge of the right wing.

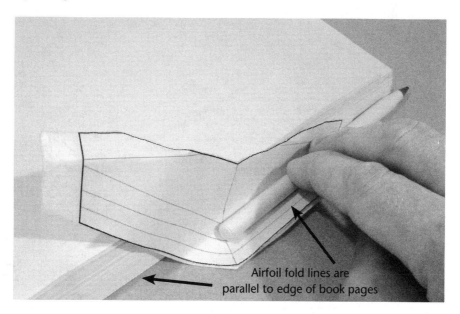

Airfoil fold lines are parallel to edge of book pages

FIGURE 6-23   Roll the pen to form a rounded fold between the first and second thin lines in from the leading edge.

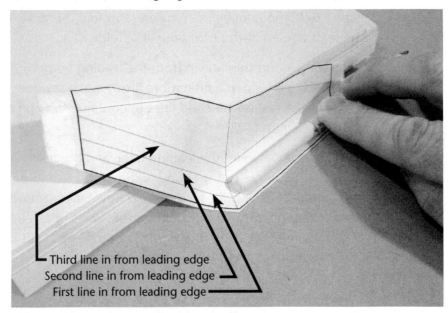

Third line in from leading edge
Second line in from leading edge
First line in from leading edge

FIGURE 6-24   Use the pen to continue the rounded airfoil fold along the right wing.

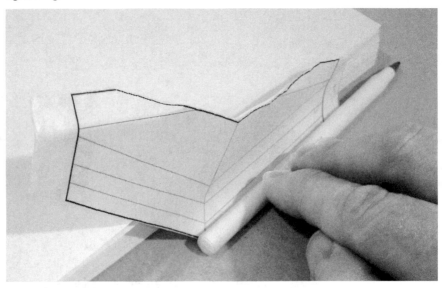

FIGURE 6-25   The finished right wing airfoil rounded fold, with a gentle angle between the second and third thin lines, becoming more angled toward the leading edge

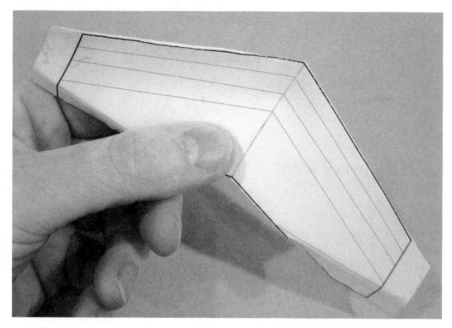

18. Create the same rolls for the left wing, so the foam slice looks like Figure 6-26.

19. This completes the folding part of the construction. Cut the template and foam along the thick lines at both wing tips to remove the template from the foam sheet, as shown in Figure 6-27.

FIGURE 6-26 Both wings have a downward rounded airfoil shape on the leading edge.

FIGURE 6-27 Remove the template from the foam by cutting the line at the wing tips.

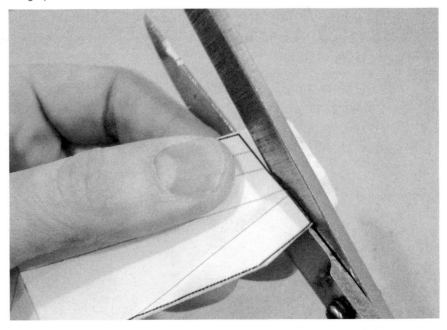

20. Break off about a fourth of a flat toothpick, and then tape the other end of the toothpick onto the centerline on the nose of the glider. Fold the tape overhanging the leading edges to the underside of the wing, as shown in Figure 6-28.

FIGURE 6-28 Tape the toothpick to the nose of the glider.

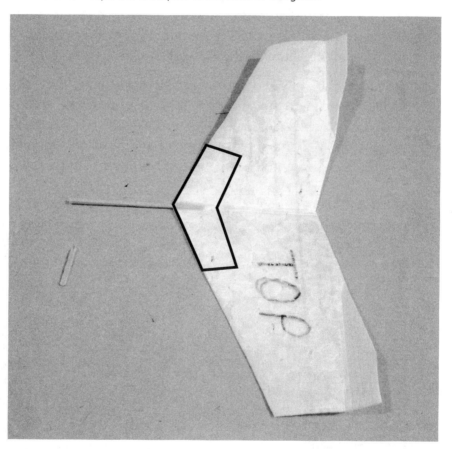

Trimming the Baby Bug

Because it also has elevons, you'll trim the baby bug similarly to how you trimmed the paper airplane surfer, X-surfer, and jumbo designs (see Table 2-1).

Start with the elevons at 45-degree angles and make a test flight, gently launching the baby bug from as high as you can at a walking speed with the nose pointed slightly down. When the baby bug is in stable flight, observe the flight path. Does it seem to dive? If so, you might need to trim off a small piece of the toothpick, as shown in Figure 6-29.

If the flight path is wavy, with alternating nose up and then down, add a little more tape to the end of the toothpick for ballast, and try again. Once the baby bug flies in a shallow, slow glide path, adjust the elevons asymmetrically to straighten any turning tendency. You might need to readjust the elevons in tandem to keep the flight path shallow.

FIGURE 6-29   If the glider dives, trim a small amount from the toothpick.

Trim a small amount of the toothpick at a time

Sustaining the Baby Bug

Once the baby bug has been trimmed to fly straight and shallow, you are ready to try flying with the paddle you made in Chapter 1. Figure 6-30 shows the flight path of the baby bug without using the paddle.

Try using the paddle to sustain the baby bug in flight. Figure 6-31 shows how the paddle can make the baby bug climb. You can also use just your hands as paddles to sustain flight, as shown in Figure 6-32.

The size of the paddle, whether your hands or the cardboard sheet, can be likened to the power of the engine on an airplane. The larger the paddle, the stronger the updraft and the better the baby bug will climb. It is much harder to sustain the baby bug with your hands!

FIGURE 6-30   The flight path of the trimmed baby bug without the paddle

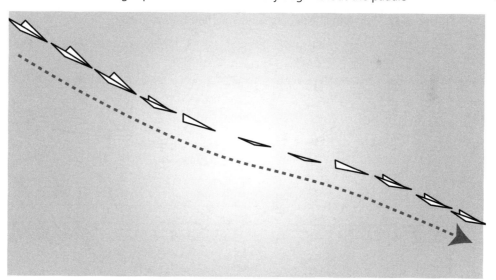

FIGURE 6-31 Placing the paddle closer to the baby bug will make it climb.

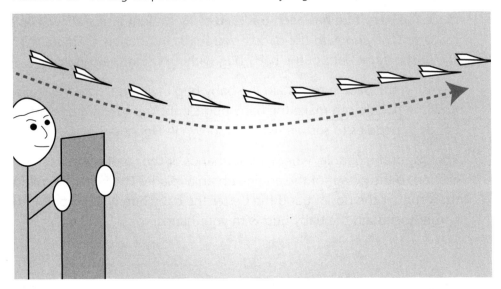

FIGURE 6-32 Flying the baby bug using your hands as a paddle

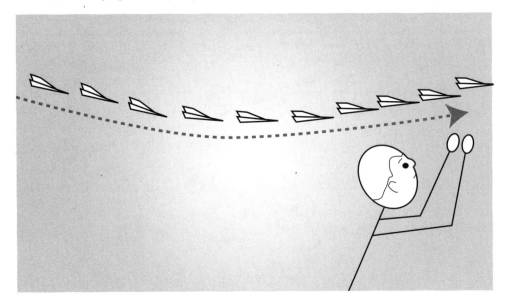

Figure 6-33 shows the baby bug flying in straight and level flight, and Figure 6-34 shows the S-turn maneuver.

FIGURE 6-33   Trajectory of the baby bug in straight and level flight

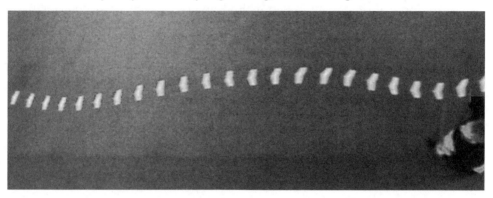

FIGURE 6-34   Trajectory of the baby bug executing an S-turn maneuver

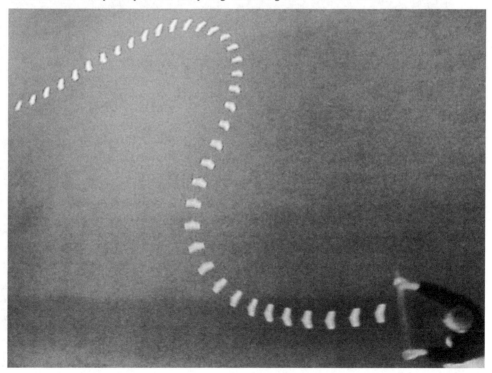

Theory

Traditional airplanes have a vertical tail fin that helps keep them flying straight. Flying wing designs such as the baby bug, X-surfer, jumbo and even the butterfly walkalong gliders do not have a vertical tail fin like a traditional airplane. What keeps the baby bug (and other flying wing designs) flying straight? Figure 6-35 shows a top view of a baby bug flying straight. In the figure the air motion is straight up. Each wing experiences the same drag from the oncoming air. The arrows in the figure mark the direction of the drag experienced by each wing; the arrows are equal in size, showing equal drag on each wing.

When the glider is flying straight, each wing experiences the same drag from the oncoming air. Figure 6-36 shows a baby bug in a skid. The wing on the right is farther into the airstream, while the wing on the left is

FIGURE 6-35   Top view of a baby bug flying straight

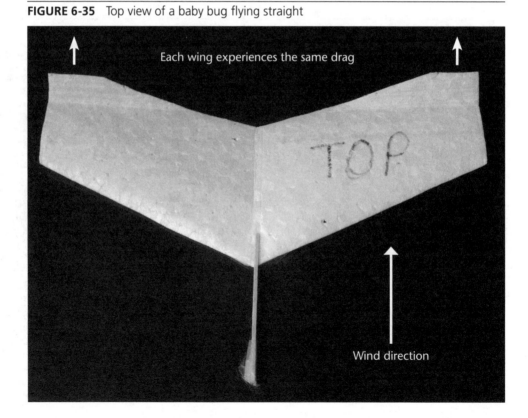

FIGURE 6-36   Top view of a baby bug in a skid relative to the oncoming air

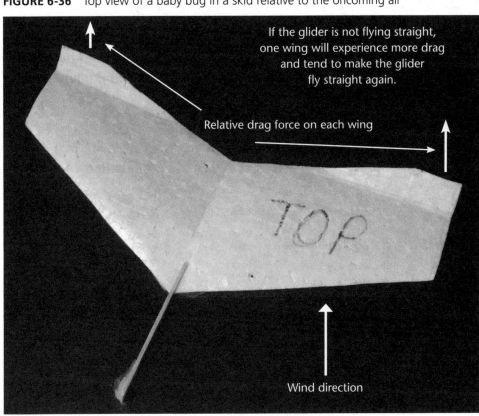

pulled inward. The different angles of each wing result in the wing on the right experiencing more drag, making the baby bug fly straight again. The arrows mark the direction of the drag experienced by each wing; the longer arrow shows more drag on the wing on the right.

Because the wings are swept back at an angle, one wing will be directed into the oncoming air more than the other, and the difference in the drag will tend to make the baby bug fly straight again.

NOTE   Both the X-surfer and the jumbo also have swept-back wings. In Chapter 4, we learned that turning the jumbo to the right causes the airplane to fly sideways, with the nose pointing to the left instead of to the right (called "adverse yaw"). The swept-back wings of the jumbo make it eventually turn in the correct direction as the turn progresses.

Hot Wire Cutter for Thin Slicing Polystyrene Foam Blocks

The instructions below describe how to build a hot wire cutter for making thin-sliced polystyrene sheets. Please consider purchasing thin-sliced polystyrene sheets from the sources described at www.walkalongglider.info before attempting the construction of a hot wire cutter.

CAUTION   _It is recommended that the hot wire cutter only be attempted with adult supervision due to the danger of a burning hot wire and toxic polystyrene fumes generated._

What you'll need:

- Expanded polystyrene foam block, 2.5-by-10-by-30 cm (1-by-4-by-12 in.).

- Recycled seafood packaging from a grocery store works well.

- Board with a smooth surface, such as the back of a clipboard, about 12.5-by-23 cm (5-by-9 in.)

- Thin cardboard (such as a cereal box)

- Scissors

- Aluminum foil

- 2 rubber bands

- 61 cm (24 in.) 40 gauge (0.0031 in. diameter) nichrome wire (available from http://jacobs-online.biz/nichrome_wire.htm)

- 2 wires with alligator clips at each end (available from Radio Shack)

- Cellophane tape

- Shipping tape

- Masking tape

- Two 9V batteries

Assembly

1.   Cut the expanded polystyrene blocks to fit the width of the board. The surface of the block should look like Figure 6-37.

2.   Cut two, 1-by-20 cm (3/8-by-8 in.) strips of thin cardboard, and tape the cardboard against the middle edges of the board, as shown in Figure 6-38.

FIGURE 6-37   The surface of an expanded polystyrene block with metric ruler for scale

FIGURE 6-38   Strips of thin cardboard are used to elevate the nichrome wire above the board's surface.

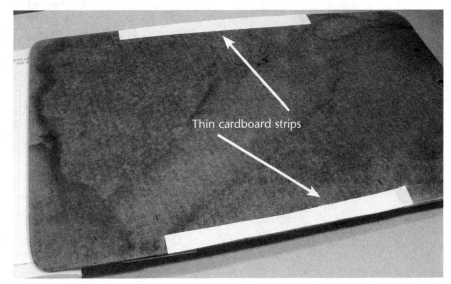

Thin cardboard strips

3. Cut two 2-by-10 cm (3/4-by-4 in.) strips of aluminum foil. Tape the strips of foil on top of the cardboard, so the edges of the foil just cover the inside edges of the cardboard (see Figure 6-39). The aluminum strips are used to make electrical contact with the nichrome wire.

4. To keep the board from burning (we'll be using electrical currents here!), line the area between the cardboard and aluminum strips with aluminum foil. Take care to leave about a 1 cm (1/2 in.) gap between the aluminum foil and the strips on either side. To prevent the polystyrene block from catching on the foil lining, tape down the foil with shipping tape along its entire length, as shown in Figure 6-40.

5. Cut each rubber band into a single strand. We'll tie knots with these bands. Start by making a loop, as shown in Figure 6-41.

FIGURE 6-39 Tape the strips of aluminum foil atop the cardboard strips.

FIGURE 6-40   Line the board with aluminum foil to prevent the hot wire from burning it. Make sure the aluminum lining does not touch either cardboard strip.

FIGURE 6-41   Make a loop with a cut rubber band.

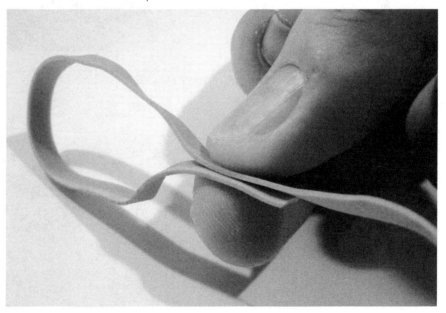

6. Wrap the loop end around your finger and tie a half hitch by tucking the loop end under itself, as shown in Figure 6-42.

7. Pull your finger out from the rubber band and tighten the knot, being sure to keep both loose ends on the opposite side of the knot from the loop.

8. Loop the end of the nichrome wire through one rubber band loop, and twist it a few times, as shown in Figure 6-43.

FIGURE 6-42   Tie off the loop by tucking it under itself, producing a half hitch knot.

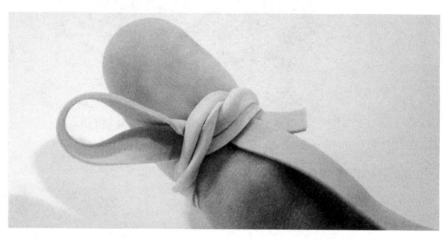

FIGURE 6-43   Attach the nichrome wire to the rubber band loop.

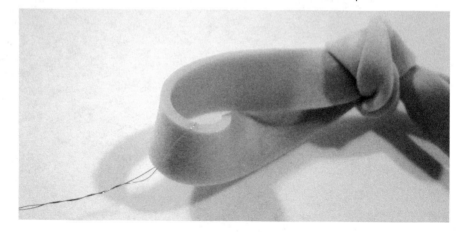

9. You'll attach both rubber band loops, loop side up, to the table top on either side of the board, as shown in Figure 6-44. Start by taping one rubber band loop (or anchor point) to the table top on one side of the board, about 15 cm (6 in.) from the midpoint of one of the cardboard strips.

FIGURE 6-44   The nichrome wire under tension across the board. Here the rubber bands are anchored underneath a secondary board, but they can also be taped to the table top with masking tape.

10. Now you'll measure and cut the nichrome wire attached to the loop. The wire should be long enough to cross the full width of the board, including the two foil strips taped to the board, with a bit of excess, which will be attached to the second rubber band anchor point taped to the other side of the table top (Figure 6-44).

11. Determine the position of the second rubber band anchor point by putting just enough tension on the wire so it runs straight across the board, as shown in Figure 6-44. Tape down the rubber band anchor point, and then attach the nichrome wire to the second rubber band loop.

12. Attach one alligator clip to the positive lead on one battery and to the aluminum strip on the side of the board (see Figure 6-45).

FIGURE 6-45   The batteries hooked up in series, with one alligator clip attached to the foil. Attaching both clips to both sides will complete the circuit and the wire will heat up.

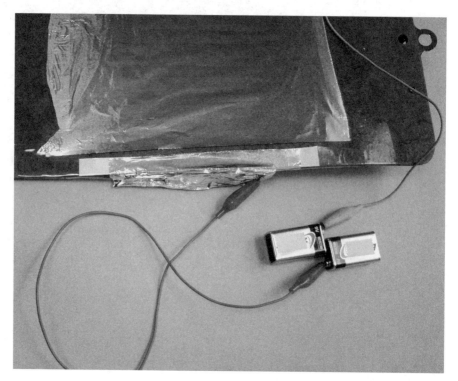

13. Attach the other alligator clip to the negative lead on the other battery.

> *CAUTION*   *Do not attach both contacts on both batteries or allow the other contacts to touch because the batteries will be destroyed in a short circuit!*

14. Attach the remaining two empty leads of the two batteries to each other to make an 18V battery (see Figure 6-45).

> *CAUTION*   *When you attach the second alligator clip to the aluminum strip, the wire will heat up quickly. Be very careful so you don't get burned!*

15. When you are ready to start slicing the polystyrene block, attach the second alligator clip to the other aluminum contact on the other side of the board, and position the block on the board, as shown in Figure 6-46.

FIGURE 6-46   When you are ready to slice the polystyrene, complete the circuit by attaching the alligator clip to the other side of the board, and then positioning the polystyrene block on the board.

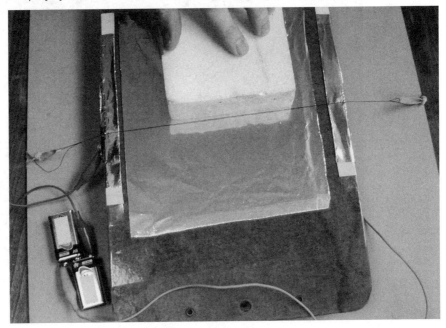

16. Move the polystyrene block so it comes in contact with the nichrome wire. Do not push too hard, because the wire will break with too much tension. In addition, if the wire is not yet hot, it will not cut the block and you will need to check the circuit again. As the wire melts into the block, gently move the block along the board, allowing time for the wire to melt the block and keeping a light tension on the wire, as shown in Figure 6-47. The finished slice is shown in Figure 6-48. Notice that this slice already has a curvature. Orient the curve so the nose and elevons are higher and separated by the dip in the curve to give the baby bug wing some washout. As soon as you are done slicing the polystyrene block, disconnect the battery.

FIGURE 6-47   Move the block gently down the block, allowing time for the wire to melt through the polystyrene and not building up too much tension in the wire.

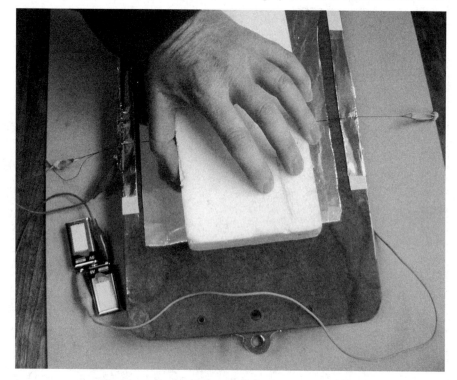

FIGURE 6-48   The thin-sliced expanded polystyrene sheet made with the hot wire slicer

7

More Fun with Walkalong Gliders

The purpose of this chapter is to explore activities that will help you become a better walkalong glider pilot and share your skills with other people. We start with planning a simple flight with the aim of landing at a predetermined location as real airplanes do. Then you can learn how to set up your own walkalong glider competition for people who have never flown as well as events aimed at more advanced walkalong glider pilots. Finally, you can take your walkalong glider show on the road, demonstrating what a walkalong glider is and teaching others in a science fair or museum environment.

Controlling Your Glider, from Takeoff to Landing

In this section, you'll learn how to keep control of your walkalong glider from takeoff to landing (see Figure 7-1).

Let's start by considering the easiest glider to fly, the tumblewing walkalong glider. To perform a controlled flight from start to finish requires planning. Consider the following:

- What will the "weather" (indoor wind) be like along the route of flight?

- Where do you want to go?

- What flight path will you take to get there?

- Where will you land?

Let's start with the "weather," or how to find calm air. Then we'll practice straight and level flight before attempting turns relative to ground reference points. The chapter finishes with how to land the glider on a table top.

FIGURE 7-1   Maneuvering a tumblewing walkalong glider to land on a table top. The white dot is a ground reference point which helps the pilot know when to turn to begin the landing pattern.

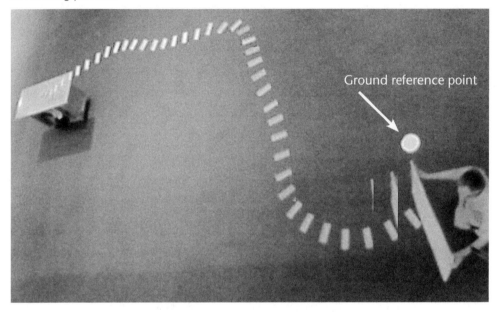

Ground reference point

Flying the Friendly Skies: Finding Calm Air

Anything that causes the air to move will affect your glider and your ability to maneuver and land where you want. The most likely sources of indoor wind are heating and cooling vents, and open windows and doors. When you fly a walkalong glider, you will notice small air movements like never before.

As with real airplanes, learning to fly a walkalong glider in smooth air is much easier. It may not be that you are a bad pilot, but that the wind is making it difficult for you to control your walkalong glider. Outdoors is generally not a good place to fly walkalong gliders unless there is little or no wind. Indoor flying gives you much more control over the sources of wind. Try flying and exploring many different indoor areas. Where are the places with calm air where it is easy to fly? Where are the bad places with turbulent air where you can barely control your walkalong glider? A good place to fly might not be where you expect it to be.

Even people walking create turbulent air movement, as shown in Figure 7-2. If a walkalong glider flies toward a person, it will appear to follow the person as he or she attempts to get out of the way. This is because the person is pulling the air behind him or her along with the glider. So steering your walkalong glider clear of the path of walking people will eliminate a source of turbulence and keep your glider in the air.

The same issue is true for you and your paddle as you are flying your walkalong glider. Doubling back on your own wake can result in loss of control in the turbulent air being pulled behind your paddle. Like a person walking by, a person flying a walkalong glider will produce turbulence, as shown in Figure 7-3.

<u>NOTE</u> *Airplanes also need to avoid each other's path because of rough air called "wake turbulence."*

FIGURE 7-2   People walking produce turbulence as the air flows in behind them, so avoid flying your walkalong glider across the path of somebody who has just walked by.

FIGURE 7-3   A person flying a walkalong glider will produce turbulence in the same way as a person walking by.

Flying Straight and Level (Harder than You Think)

If you can fly straight and level with no changes in direction or height, you are truly blessed with calm air. More likely, attempting to fly straight and level will reveal any slight wind in your flight path. Figure 7-4 shows a straight-and-level flight in relatively still air.

FIGURE 7-4   Flying straight and level will reveal any wind currents present. The two white dots are ground reference points to help the pilot center his path within the camera frame.

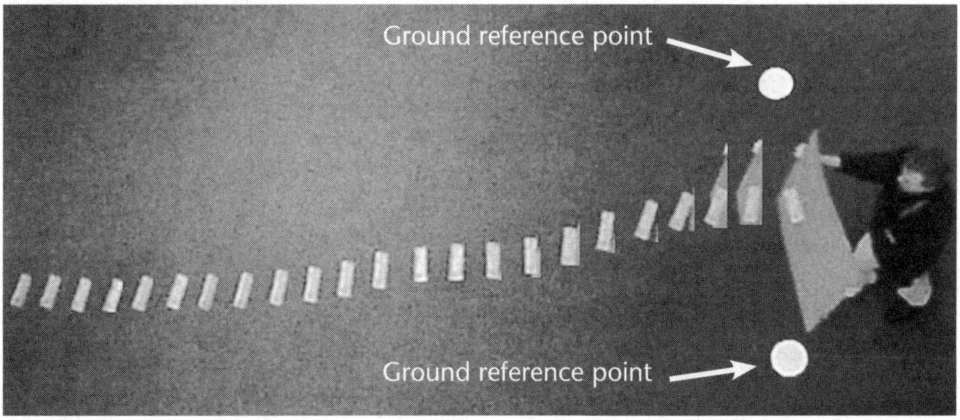

Ground reference point

Ground reference point

Don't get discouraged if you're having trouble, because real airplanes get tossed around on windy days, too. If you keep your glider flying high, such as above eye level, it will be easier to regain control if your glider gets off course.

Approaching the Landing Site (Turns Relative to Ground Markers)

Two steps are involved in landing: first, approaching the landing site, which will require straight and level flight, and, second, descending to the landing and touchdown. To practice approaching the landing site, we'll direct the glider to fly between two markers on the ground, as shown in Figure 7-5. This maneuver would be like flying through a doorway into another room.

If you wanted to fly though a doorway, for example, you would need to stay as far away from the walls and the doorjamb as possible. You will need an extra distance margin in case any wind throws you off course. Orienting your flight path perpendicular to the doorway will give you maximum clearance on both sides (especially if you approach the doorway from any direction other than perpendicular).

FIGURE 7-5 This flight trajectory involves turns relative to ground markers (white dots).

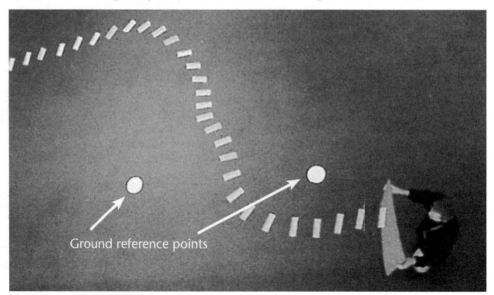

Ground reference points

In flying relative to ground objects, you are applying the skills you learned in earlier chapters about how to steer a walkalong glider and going a step further to maneuver relative to obstacles. Being able to turn your glider to fly between the markers will help you when you turn from the base leg to the final approach when landing.

<u>NOTE</u>   *Over-shooting the turn to final approach is one of the most critical mistakes in landing real airplanes.*

Landing Your Walkalong Glider

Flying an airplane to arrive safely on the ground requires choosing a touchdown point and a way of getting there, known as a "landing pattern." Airplane landing patterns can be quite complicated, but our walkalong glider landing pattern will involve one or two segments. We'll start with just a straight-in descending leg called a "final approach," where our flight path will already be aligned with the touchdown area.

Using a table or other elevated surface as your touchdown area and landing spot is helpful, because you can use your paddle to control the flight right up to the edge of the table. (If you use the floor, the paddle would probably scrape on the ground long before the glider lands on the floor.)

It is better to land on a table top lengthwise than using the shorter width; landing lengthwise allows more room for undershooting or overshooting and still landing on the table top. Arriving at the table top from a direction other than the final approach direction (aligned lengthwise with the table top) will require another leg, or a "base leg," flown perpendicular to the "final approach leg" (see Figure 7-6). Each leg involves a different adjustment. The base leg is flown until the glider is aligned with the landing site. The final approach leg is a straight descending flight path arriving at the table-top height.

The base leg involves straight flight, keeping an eye on the runway, and judging when to turn so the walkalong glider will be headed toward the runway on final approach. Final approach involves descending while keeping the glider flying straight toward the touchdown point.

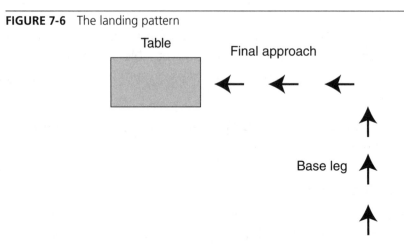

FIGURE 7-6 The landing pattern

Start by flying a final approach to your landing site. Try and get a feel for how far away you need to be so you can still steer the glider and descend slowly. Starting farther away from the landing site will allow you to make the glider descend more slowly and makes steering the glider easier. This is because the paddle produces lift to steer the glider, and a "free glide" (with a maximum descent rate) would mean you'd have little or no control. It might take you several tries to get the feel of the way the glider steers while descending.

Once you can get the glider to land on the table, try starting on a base leg, so you'll turn the glider into final approach. Once you are successful at the base leg and final approach, you will know where you need to begin the landing phase of your flights. You will find it helpful, as do real pilots, to keep your landing patterns as similar as possible, leaving yourself time and space to adjust for wind.

Walkalong Glider Competitions

This section is divided in two parts. The first outlines competition events geared toward newcomers to walkalong gliding. The second parts offers a series of advanced competition events, some based on great moments in aviation history. These imaginative events were developed by David Aronstein, an aerospace engineer and walkalong glider enthusiast.

Newcomer-Oriented Walkalong Glider Competitions

Walkalong gliding is new to most people, and most walkalong glider competitions are geared toward newcomers. Most competitions are held in a large space, such as a gym. Half of the gym or a large indoor area is often set aside for a practice area, where people become acquainted with flying walkalong gliders for the first time and contestants can improve their flying skills; the other half is used for walkalong glider competition.

Launch Helpers

Launch helpers can launch contestants' gliders both in the practice area and in competitions if contestants opt for help. Using a launch helper allows contestants to concentrate on flying and using the paddle. The launch helper holds the glider high above and forward of the paddle. The contestant holds the paddle, ready to fly when the glider is released. The launch helper then walks together with the contestant (Figure 7-7), getting the glider up to its proper speed. (In the case of the tumblewing, the

FIGURE 7-7   The launch helper gets the walkalong glider flying smoothly so the pilot can concentrate on sustaining and controlling the glider in flight.

launch helper doesn't need to move because of its slow launch speed.) The launch helper must also release the glider at the proper angle and speed so that the glider is in steady flight in the ridge lift generated by the paddle. Bear in mind that the launch helper guides the glider up to airspeed, and not the other way around. Most beginners swap between flying and being a launch helper.

Types of Competitions

Simple walkalong glider competitions involve events of increasing difficulty. Each event should have a separate class for each type of glider (for example tumblewing, X-surfer, or paper airplane classes). For newcomers, generally the slower the glider flies, the easier it is to fly. Slower speeds also reduce paddle turbulence and the air is generally smoother and easier to fly in.

The Duration Event

The duration event times how long each contestant can keep the walkalong glider flying from launch to when the glider makes contact with the ground. The duration event does not require that the pilot be able to steer the glider.

As soon as a pilot is able to sustain a walkalong glider, he or she is ready for the duration event. How long can you keep your walkalong glider flying? An important skill to master for the duration event is making the glider fly higher to recover better from temporary loss of control. (Flying higher is also good practice for the altitude event, discussed a bit later.)

Depending on the speed of the walkalong gliders being flown, the competition area can be divided into separate zones to time multiple contestants' durations simultaneously. It is best to have a judge timing each contestant, even if they are all started at the same time.

The Distance Event

The distance competition will involve launch from behind a start line and a flight path which must round reference points a given distance apart. Several contestants can compete simultaneously using the lane configuration shown in Figure 7-8.

FIGURE 7-8 Setup for a distance competition event, in which four contestants can compete simultaneously

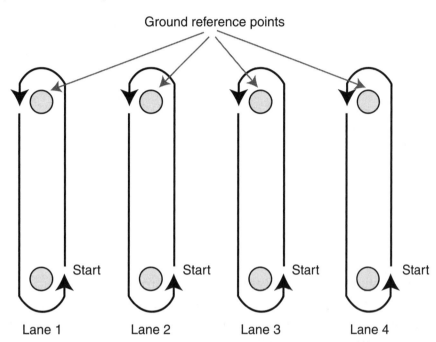

Ground reference points

Lane 1   Lane 2   Lane 3   Lane 4

The circles in the figure denote cones or other ground reference points that are spaced to provide generous area for turning at each end. The winners will have flown the most number of laps (plus the fraction of a lap as determined by the judges) before the glider contacts the ground or the contestant misses or hits one of the points. Each flight can also be timed as in the duration competition, so duration records may also be broken in the distance competition.

The Altitude Event

The altitude event will involve launching from behind a gate, 6 meters (20 ft.) from a level string at a starting height of 1 meter (3 ft.). The contestant will attempt to gain altitude to fly the glider above the string (see Figure 7-9). The contestant launches the glider from behind the start gate and attempts to make the glider fly high enough to fly over the string.

FIGURE 7-9 Setup for height contest

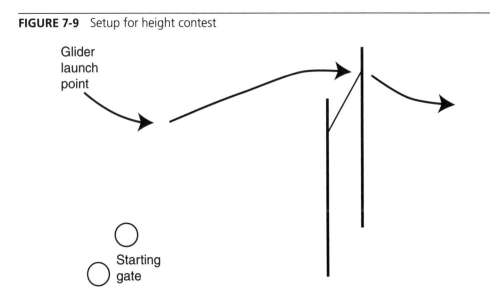

The height of the string is raised with each round, and as long as at least one contestant succeeds in flying above the string, each contestant can attempt the higher string, even if they failed to achieve a lower height. Contact with the string (pilot, paddle, or glider) does not count. Judges should be positioned to sight down the string to verify that the glider successfully clears the string.

Advanced Walkalong Glider Competitions

The following walkalong glider competition events were created by David Aronstein and are based on some of aviation's greatest feats.

The Airlift Event

The airlift event involves flying a single glider across an obstacle (here a "river") and back as many times as possible in one minute (see Figure 7-10).

An airlift is the movement of supplies or people mainly by air. The airlifts of note in aviation history are the "Hump" operation and the Berlin Airlift, both of which occurred during World War II. The Hump occurred

FIGURE 7-10   Diagram of the airlift event, in which two pilots must keep a glider flying, passing it to each other across the river

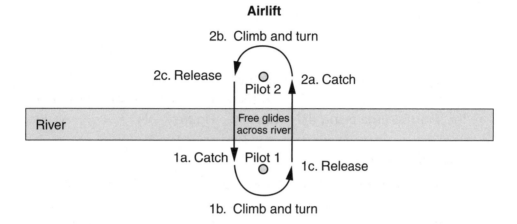

between India, Burma, and China from 1942 to 1945 across the Himalaya Mountains and overcame the blockading of China by Japanese forces. The Berlin Airlift, from June 1948 to September 1949, overcame the blockading of the city of Berlin by Soviet forces. In each of these airlifts, time was of the essence to move as many supplies as possible.

In the walkalong glider airlift event, the team with the most crossings in a minute wins. Teams must keep the glider flying at all times, and when the glider makes contact with the ground, the team will be scored only for the crossings up to when the glider landed.

David Aronstein offers these tips for doing well in the airlift event (from a letter sent to contestants before a walkalong glider competition):

- For the catch—stand to one side of the incoming glider's path—*not* directly in front of it!

- For the release—*do not* stop the board suddenly. This will send an air disturbance forward, which will overtake the glider!

- You want to be able to receive it, turn it around while *gaining* all the altitude necessary for the return trip, and release it smoothly in the general direction of your partner.

- Although it is a timed event, remember the "3 C's": Calm, Cool, and Consistent.

- Practice your climbing turns.

The Pylon Race

The pylon race is flown around a triangular course and is based on the Air Races held in Reno, Nevada, and elsewhere. Pilots fly two laps around, with the shortest time being the winner (see Figure 7-11).

The Obstacle Course

This race involves a number of tasks, including but not limited to flying through hoops, through gates, over tall objects, and under limbo sticks. In the case of the hoop, the pilot must not only successfully maneuver the plane through but also catch the plane as it exits the hoop to continue flying. The obstacle course shown in Figure 7-12 involves flying through two hoops and around a four-cone slalom, and finishes with a table-top landing.

FIGURE 7-11 Pylon race course setup, in which pilots are timed over two laps around the course

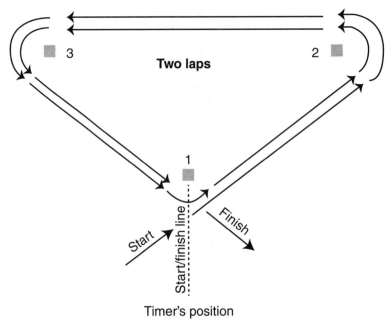

FIGURE 7-12   Obstacle course involving flying through two hoops, around a four-cone slalom, and finishing with a table-top landing

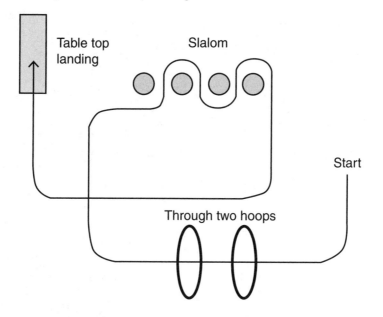

Aerial Jousting (Dog Fighting)

In an aerial jousting or dog-fighting event, participants attempt to fire pulses of air using their paddles to dislodge other gliders while keeping their own walkalong glider flying and out of trouble. The winner is the last person left flying. Figure 7-13 shows two walkalong glider pilots sparring in an aerial joust.

This event is based on the battle for air supremacy of World War I, when flying aces such as the Red Baron (Manfred von Richthofen) ruled the skies.

Ideas for Presenting Walkalong Gliding

A walkalong gliding presentation provides an attention-grabbing activity for venues such as science fairs and museum environments. A walkalong glider demonstration naturally lends itself to the audience or passersby trying it for themselves or even building and flying a design from materials on hand.

FIGURE 7-13 Two walkalong glider pilots try and dislodge each other's gliders while keeping their own flying in an aerial joust or dog-fighting event.

People will experience not only controlling a simple airplane in flight but also an interactive hands-on demonstration of soaring flight.

Making Supplies for Building Tumblewings

Once you have found a source of light paper, it can be cut to proper size using a paper cutter, as shown in Figure 7-14. Note the long dimension of the tumblewing is oriented up and down, parallel to the grain of the paper, as shown in Experiment 1 in the Theory section of Chapter 2.

The large blade and ruler are perfect for sizing multiple sheets at once. The ratio of length to width of the paper strips should be between 1:3 and 1:4.

FIGURE 7-14   Use a paper cutter to size multiple sheets of paper for tumblewing construction.

Thus, if the sheets are of dimensions 22-by-28 cm (8 1/2-by-11 in.), cut 4 cm (1 1/2 in.) strips lengthwise and cut each 22 cm (8 1/2 in.) strip in half (see Figure 7-15) for strips of dimension 14-by-4 cm (5 1/2-by-1 1/2 in.).

The resulting stack of paper strips can be easily stored in an eyeglass case (see Figure 7-16) that fits in your pocket.

Getting People Interested

As with all walkalong glider flying, finding still air and a wide open area are important to giving an impressive demonstration and making it easier for people to try walkalong gliding. As mentioned earlier, people walking will make areas of turbulent air that are easier to avoid in a large area. Using a slow glider such as the tumblewing makes the demonstration easier and above all safer.

FIGURE 7-15   Cut the paper strips widthwise.

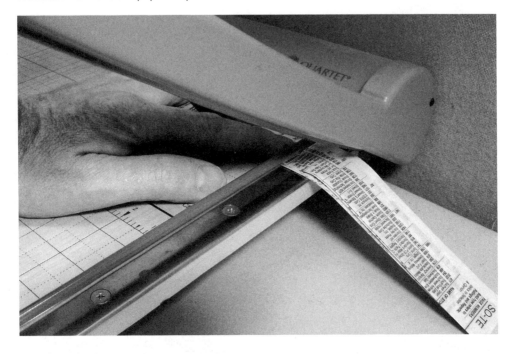

FIGURE 7-16   A stack of paper strips for constructing tumblewings stored in an eyeglass case

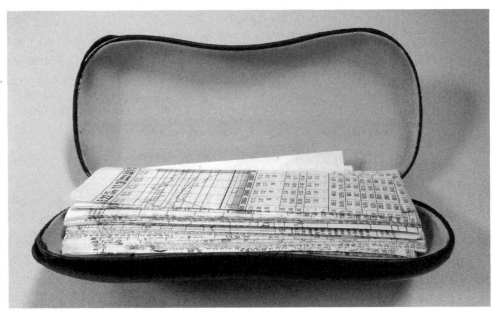

I usually fly my glider around the room to catch people's attention. Once I get a question such as What is that?, How does it work? or Can I try that?, I try and challenge them to guess what is keeping the plane flying: is it electrical, magnetic, or air (some kind of updraft)? I get many different answers from a group of people. When somebody says "wind" or "air," they have to be supported and asked to figure out how the air moving around the paddle keeps the glider flying. I find that when people understand how walkalong gliding works, they find it much easier to learn to fly.

Walkalong Glider Flying Lessons

I do the walkalong glider demonstration with one paddle to minimize turbulence and focus attention in one place. When people want to try flying the glider, I launch the glider for them so that they can concentrate on flying. Even smaller children can participate by holding the paddle as the glider flies. If I catch the attention of a group of people, I ask them to make a line extending back from the launch point, like a busy airport where the airplanes are all lined up waiting to take off.

Before releasing the glider, I tell the person with the paddle to follow the glider, and that when the glider descends to the level at the top edge of the paddle, go a little faster, as if to run over the glider. Then again after launching the glider, as it glides down to the top edge of the paddle, I urge the person to go faster. When the glider stops descending, I congratulate them on a successful first flight.

For the second walkalong glider flying lesson, I show the person how to steer the glider. In the case of the tumblewing, angling the paddle relative to the tumblewing glider, as shown in Chapter 1, will make the glider turn. Oddly, keeping the glider going in a straight line is a good exercise for steering the glider. This is because, more often than not, the glider will tend to turn on its own or will turn because of drafts in the room. Flying an airplane from point A to B shows more control than flying aimlessly or around in circles. I also urge the person to keep the paddle angled relative to the glider until the glider turns (have patience). Walkalong glider presentations allow a diverse and especially younger group of people to safely experience controlling an airplane in flight.

Demonstrating Walkalong Glider Construction

For people interested in building their own tumblewing walkalong gliders, I distribute the paper strips I cut ahead of time. Before folding, I ask people to note how floppy the paper is.

Then I ask them to start by folding the two long folds, as shown in Chapter 1. After the first long fold, I draw their attention to the rigidity that the fold gives to the paper. After folding just the two long folds, I ask them to drop the glider to see how it flies. Without the winglets, the tumblewing will tend to fly sideways, similar to the right-hand flight trajectory in Chapter 1. Then they complete the glider by adding the winglets, and I show how the winglets make the tumblewing glider resistant to flying sideways. I urge people to take their gliders home and use a pizza box as a paddle.

Conclusion

I hope this book has given you an appreciation for life on the wing, human-powered flight, and how much effort and coordination you need to fly even the lightest model walkalong gliders. This book is just the beginning. You have the potential to contribute new ideas to the field of walkalong gliding. There are many improvements that would make for bigger and better glider designs, such as lighter materials and structures. Think of the butterfly, so light and yet strong.

Glossary

adverse yaw
When entering or exiting a turn, the airplane briefly angles in a direction opposite to, or adverse to, the desired turn.

aileron
A movable surface on the trailing edge of airplane wings that allows the plane to be controlled about the roll axis for performing turns.

airfoil
The shape of a wing in cross-section. An airfoil-shaped body is optimized (minimize drag and maximize lift) for producing an aerodynamic force when moving through a fluid such as air.

airlift
Organized delivery of people and/or goods mostly by aircraft.

airplane
Vehicle capable of flight by moving forward through the air.

altitude
Height above ground.

ballast mass
Mass positioned aboard an aircraft specifically to adjust the aircraft's center of gravity.

boom
A spar, pole, or extension used to increase the moment of a mass, usually for adjusting the center of gravity of a glider while minimizing the increase in weight needed.

center of gravity
A point in space at which the entire mass of an aircraft balances. This concept was developed by Archimedes around 200 BCE and greatly simplifies calculations.

Depron foam
Styrofoam sheet suitable for building model aircraft.

descent rate
Vertical speed or a downward change in altitude.

dihedral
Upward angle of an aircraft's wings from horizontal. This upward angle to the wings makes an aircraft more stable about the roll axis.

drag
Force acting opposite the direction of movement. Drag is a force caused by air flowing past the airplane's surface and is the component parallel to the air motion. For an airplane in stable flight, drag is equal and opposite to the thrust force (see Figure G-1).

FIGURE G-1   Forces acting on an airplane in stable flight

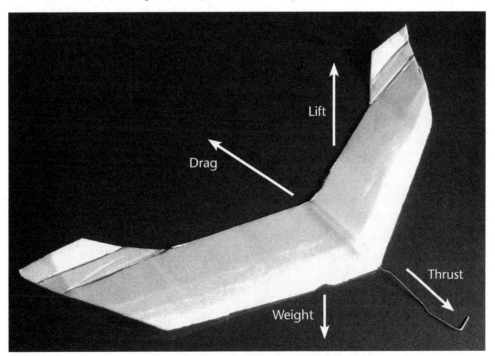

elevator
Movable surfaces on the trailing edge of the horizontal stabilizer of an airplane used to move the nose up or down and change the angle of the wings. The elevator is responsible for control about the pitch axis.

elevon
Movable surface that combines the function of the elevator and ailerons on flying wing aircraft designs.

final approach
The last leg of flight before an airplane lands. This leg is aligned with the runway.

flying wing
A tailless fixed-wing airplane with no definite fuselage.

foam core
Sheet material consisting of an inner layer of polystyrene sandwiched between paper facing.

inboard
The area toward the centerline of an aircraft.

landing pattern
Standard path followed by airplanes before landing at an airport.

lift
Force responsible for keeping an airplane in the air. Lift is a force caused by air flowing past the airplane's surface. Lift is the component vertical and perpendicular to the air's motion. For an airplane in stable flight, lift is equal and opposite to the weight of the airplane (see Figure G-1).

nose
The frontmost part of an airplane.

outboard
Away from the centerline of an aircraft.

paddle
Tool used for pushing against the air to produce an updraft for sustaining and controlling a walkalong glider.

paper grain
The direction of fibers that make paper stronger on one direction, usually oriented parallel to the longer dimension of the sheet.

pitch axis
One of the aircraft principal axes of rotation, which is horizontal and perpendicular to the motion of the airplane (see Figure G-2). The pitch axis is responsible for moving an airplane's nose up and down.

FIGURE G-2   The pitch axis

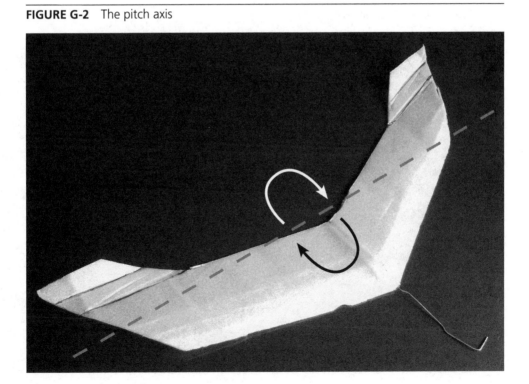

roll axis

One of the aircraft principal axes of rotation, which is parallel to the motion of the airplane (see Figure G-3). The roll axis is responsible for moving an airplane's wingtips up and down.

science fair

A competition in which contestants display a report, poster board, and models describing a science project.

soaring flight

Gliding in a rising air current. Although the glider is falling relative to the air, the air is going up faster than the glider comes down.

speed brake

Surfaces on an airplane designed to increase drag.

FIGURE G-3   The roll axis

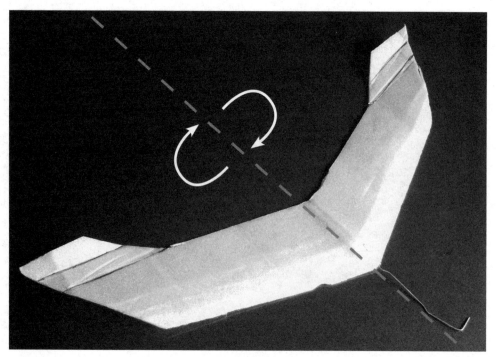

stall
Reduction in lift of an airfoil when the airfoil is angled too much relative to the oncoming air flow.

tail
The assembly of surfaces on the rear end of most airplanes; used for stability.

template
An overlay used to replicate shapes or designs.

thermoform
A manufacturing process by which a plastic is heated to a pliable temperature and formed to a specific shape.

trailing edge
The rear part of an aerodynamic surface such as a wing.

trajectory
The flight path an airplane or other object follows through the air as a function of time.

trim
The process of adjusting the aerodynamic surfaces of an airplane or glider to optimize its flight characteristics.

tumblewing
A paper airplane design that rotates about a horizontal axis, perpendicular to the direction of flight.

turbulence
Air flow characterized by chaotic wind directions.

updraft
A vertical wind blowing upward.

walkalong glider
A lightweight slow-flying model aircraft designed to be kept aloft in the updraft created by a paddle held by the pilot walking along with the aircraft.

wing loading
The loaded weight of an aircraft divided by the area of its wings.

wing root
The part of the wing closest to the centerline of an airplane. The opposite end of a wing is called the wing tip.

wing washout angle
A twist in a wing that allows for a higher angle of attack at the wing root and a lesser angle at the wing tip. Wing washout is a design feature on airplanes that makes the wing stall first at the wing roots before the wing tips and helps the wings remain level in a stall.

winglet
A wing tip device. The winglets on a tumblewing prevent the tumblewing from sliding sideways.

wing tip
The end of the wing farthest from the airplane centerline. The wing tip is on the opposite end of the wing from the wing root.

yaw axis
One of the aircraft's principal axes of rotation, which is vertical and perpendicular to the motion of the airplane (see Figure G-4).

FIGURE G-4   The yaw axis allows an airplane's nose to move side to side.

Index